10 Springer Series in Solid-State Sciences
Edited by Manuel Cardona and Peter Fulde

Springer Series in Solid-State Sciences

Editors: M. Cardona   P. Fulde   H.-J. Queisser

Volume 1   **Principles of Magnetic Resonance**   2nd Edition
By C. P. Slichter

Volume 2   **Introduction to Solid-State Theory**
By O. Madelung

Volume 3   **Dynamical Scattering of X-Rays in Crystals**
By Z. G. Pinsker

Volume 4   **Inelastic Electron Tunneling Spectroscopy**
Editor: T. Wolfram

Volume 5   **Fundamentals of Crystal Growth I.** Macroscopic Equilibrium and Transport Concepts
By F. Rosenberger

Volume 6   **Magnetic Flux Structures in Superconductors**
By R. P. Huebener

Volume 7   **Green's Functions in Quantum Physics**
By E. N. Economou

Volume 8   **Solitons and Condensed Matter Physics**
Editors: A. R. Bishop and T. Schneider

Volume 9   **Photoferroelectrics**
By V. M. Fridkin

Volume 10   **Phonon Dispersion Relations in Insulators**
By H. Bilz and W. Kress

Volume 11   **Electron Transport in Compound Semiconductors**
By B. R. Nag

Volume 12   **The Physics of Elementary Excitations**
By S. Nakajima

Volume 13   **The Physics of Selenium and Tellurium**
Editors: E. Gerlach and P. Grosse

Volume 14   **Magnetic Bubbles**
By A. H. Eschenfelder

Volume 15   **Modern Crystallography I.** Crystal Symmetry, Methods of Structural Crystallography
By B. K. Vainshtein

H. Bilz   W. Kress

Phonon Dispersion Relations in Insulators

With 162 Figures in 271 Separate Illustrations

Springer-Verlag Berlin Heidelberg New York 1979

Professor Dr. Heinz Bilz
Dr. Winfried Kress

Max-Planck-Institut für Festkörperforschung
Heisenbergstrasse 1, D-7000 Stuttgart 80, Fed. Rep. of Germany

Series Editors:

Professor Dr. Manuel Cardona
Professor Dr. Peter Fulde
Professor Dr. Hans-Joachim Queisser

Max-Planck-Institut für Festkörperforschung
Heisenbergstrasse 1, D-7000 Stuttgart 80, Fed. Rep. of Germany

ISBN 3-540-09399-0 Springer-Verlag Berlin Heidelberg New York
ISBN 0-387-09399-0 Springer-Verlag New York Heideberg Berlin

Library of Congress Cataloging in Publication Data. Bilz, Heinz, 1926-. Phonon dispersion relations in insulators. (Springer series in solid-state sciences ; 10). Bibliography: p. Includes index. 1. Crystals, Effect of radiation on--Charts, diagrams, etc. 2. Phonons--Scattering--Charts, diagrams, etc. I. Kress, Winfried, 1941- joint author. II. Title. III. Series. QC176.8.R3B54   530.4'1   79-17304

This work is subject to copyright. All rights are reserved, whether the whole or part of the material is concerned, specifically those of translation, reprinting, re-use of illustrations, broadcasting, reproduction by photocopying machine or similar means, and storage in data banks. Under § 54 of the German Copyright Law, where copies are made for other than private use, a fee is payable to the publisher, the amount of the fee to be determined by agreement with the publisher.

© by Springer-Verlag Berlin Heidelberg 1979
Printed in Germany

The use of registered names, trademarks, etc. in this publication does not imply, even in the absence of a specific statement, that such names are exempt from the relevant protective laws and regulations and therefore free for general use.

Offset printing: Beltz Offsetdruck, Hemsbach/Bergstr. Bookbinding: J. Schäffer oHG, Grünstadt.
2153/3130-5 4 3 2 1 0

Preface

This phonon atlas presents a collection of phonon-dispersion and density-of-states curves of more than a hundred insulating crystals. It grew out of an appendix to a handbook article on phonon spectra [2.1] from which it was finally separated mainly because this phonon atlas provides a rather self-contained tool for every scientist who is working in the field of dynamical properties of solids. He often may find it useful to have a handy documentation of the experimental phonon dispersion curves which have been measured so far, together with information on calculated dispersion relations and densities of states.

The book will be found to be incomplete by readers who are interested not only in phonon frequencies of a specific crystal but would also like to know about related properties such as elastic and dielectric constants. This is, at the present time, beyond the scope of this volume, but the authors would welcome all suggestions and criticism which could be considered for a forthcoming edition. Furthermore, we would be pleased to provide interested readers with information about phonon spectra which came to our knowledge after completion of the manuscript. On the other hand, we will be most grateful for all information about phonon dispersion curves which is missing in our collection or new data for further editions.

During the early stage of this work, Dr. Erika Kiefer-Schröder and Professor Ulrich Schröder gave us important help in the setup of this documentation and its realization. The authors are most grateful for their contributions. In addition, the authors have been encouraged by many colleagues to undertake this attempt of phonon documentation. We are indebted to all of them. The FIZ 4 at Karlsruhe generously provided us with information retrieval runs. We thank them, in particular Dr. Behrends, Dr. Ebel, and Dipl. Phys. E. Zybell, for their support. One of the authors gratefully acknowledges the financial support by BMFT, Bonn. Finally, it is a real pleasure to thank Miss Gisela Keck for her careful and patient preparation of the manuscript.

Stuttgart, September 1979　　　　　　　　　　　　　　　　H. Bilz　　W. Kress

Contents

Part I. Summary of Theory of Phonons

1. Introduction 3
2. Phonon Dispersion Relations and Phonon Models 6
   2.1 Formal Force Constants (FCM) 6
   2.2 Rigid Ion Model (RIM) 7
   2.3 Dipole Models (DM) 8
       2.3.1 Shell Model (SM) 8
       2.3.2 Extended Shell Model (ESM) 10
       2.3.3 Overlap Shell Model (OSM) 10
       2.3.4 Deformation Dipole Model (DDM) 10
       2.3.5 Deformable Shell Model (DEFSM) 11
       2.3.6 Breathing Shell Model (BSM) and Double Shell
             Model (DSM) 11
       2.3.7 Quadrupole Shell Model (QSM) 12
       2.3.8 Bond Charge Model (BCM) 12
       2.3.9 Valence Force Field Model (VFFM) 12
       2.3.10 Valence-Overlap Shell Model (VSM and VOSM) 13
       2.3.11 Three-Body-Force Shell Model (TSM) 13
   2.4 Microscopic Theory and Models 14
   2.5 Abbreviations Used in Figure Captions 15
   2.6 Conversion Factors Between Units Used to Express
       Phonon Energies 16

Part II. Phonon Atlas of Dispersion Curves and Densities of States

3. Rare-Gas Crystals 19
4. Alkali Halides (Rock Salt Structure) 27
5. Metal Oxides (Rock Salt Structure) 49
6. Transition Metal Compounds (Rock Salt Structure) 59
7. Other Cubic Crystals (Rock Salt Structure) 73
8. Cesium Chloride Structure Crystals 85

9. Diamond Structure Crystals	95
10. Zinc-Blende Structure Crystals	101
11. Wurtzite Structure Crystals	117
12. Fluorite Structure Crystals	123
13. Rutile Structure Crystals	133
14. ABO_3 and ABX_3 Crystals	141
15. Layered Structure Crystals	161
16. Other Low-Symmetry Crystals	179
17. Molecular Crystals	193
18. Mixed Crystals	203
19. Organic Crystals	209
References	219
Subject Index	231

Part I

Summary of Theory of Phonons

1. Introduction

One of the most powerful methods to determine phonon dispersion relations in a crystal is inelastic neutron scattering which yields phonon frequencies of all wavevectors with an accuracy of a few percent or better if circumstances are favorable. Infrared and Raman spectra provide optic frequencies only at very long wavelengths with an accuracy of about 0.1%. From these data, and including information obtained from elastic and dielectric constants, it is possible in many cases to calculate phonon dispersion relations which interpolate between the experimental data. In the simplest case, a set of interionic force constants is fitted, usually by a least-square procedure, to the data in such a way as to give agreement within the limits of experimental error. The number of force constants which is required in such a procedure becomes, however, often rather high (≥ 20) even in simple cubic crystals. Therefore, phonon models are widely used which start from a parametrized "model" description of the electron-ion interaction before going, via the adiabatic condition, to the "formal" force constants. This leads usually to a drastic reduction of parameters (<10 for diatomic crystals) and facilitates considerably the physical interpretation of the phonon spectra. A summary of the most important models and a short discussion of the underlying physical ideas is given in Part I.

In Part II dispersion curves and densities of states of more than 100 insulating crystals are represented. They are grouped into families mainly determined by the crystal structure. In some cases a separate chapter is devoted to a specific group of crystals, e.g., the alkali halides, which has been particularly well investigated. The transition metal compounds with rock salt structure which are in fact metals, are included (Chap. 6) since they exhibit many interesting features related to insulating crystals with rock salt structure although they show characteristic differences, too. At the beginning of each chapter, the individual crystals of the family are listed, together with a reference to the number of the figure in the text. In addition an alphabetic list of crystals is given in the subject index at the end of the book. The table on p.15 informs the reader about abbreviations used in the figure captions. These captions contain, in a very condensed form, the necessary information about literature, thermodynamic data (temperature and pressure), the type of phonon model, and the number of parameters used in the calculation. The short notations under M: FCM, RIM etc. refer to corresponding subsections in Part I. Additional information can be found in the theoretical papers cited in the figure captions. For the majority of crystals, usually only room-temperature data are available. In a few cases, low-temperature measurements have been carried out and are reported in the text. In cases where new data have replaced less accurate older measurements, only the new results are represented; reference to the former data is usually made in the literature cited.

While selecting the dispersion curves of a crystal for the atlas, emphasis was put on the original experimental papers even if, occasionally, slightly better calculations were available. These are cited in the figure captions. In case of the densities-of-states curves, the most reliable model calculation was chosen except for the very few cases where direct measurements existed. The number of references cited for a given crystal was limited in order to prevent captions from becoming too cumbersome.

The decision about the theoretical calculations to be reported on in the text occasionally gave the authors a splitting headache because of conflicting arguments to be considered. On one side, the scientist usually looks for a description of data which is as precise as possible so that related entities, in particular densities of states, can be determined with high accuracy. On the other side, the theoretical models used in these types of careful fitting-to-the-data calculations are not always the best from a physical point of view. Sometimes, they contain a certain number of parameters which prevent useful extensions of the model to other entities such as Raman scattering, etc. We have decided in several cases where a simpler and "better" model (in the author's opinion) was nearly as good as a more accurate one to show the first version. There is always sufficient literature given in the captions to find out about further theoretical calculations. In one case (ZnO) the situation was found to be extremely controversial; therefore, two different versions are presented in the text. The authors hope that some "white spots" in this atlas may stimulate further experimental investigations and, quite generally, that low-temperature and high pressure data will become available to a larger extent during the coming decade.

It is felt that many of the users of this volume (aside from the few experts in the field) would welcome a short introduction to the different models; this is given in the following chapter. For more details, the reader is referred to the many textbooks and articles in the field. A selection is given in the references. We note that notation always follows closely the above-cited handbook article on phonon spectra [2.1].

2. Phonon Dispersion Relations and Phonon Models

2.1 Formal Force Constants (FCM)

The lattice potential ϕ of a crystal in thermal equilibrium can be derived from the free energy $F(T,V)$. ϕ reads in the classical harmonic approximation ($\phi \simeq \phi_2$),

$$\phi_2 = \frac{1}{2} \sum_{L,L'} \underline{u}(L) \underline{\phi}(LL') \underline{u}(L') \tag{2.1}$$

where $L = (\ell,\kappa)$ denotes the κth particle in the ℓth cell. The $\underline{u}(\ell,\kappa)$ denote the displacement vectors of the ions with equilibrium positions, $\underline{x}(\ell,\kappa) = \underline{x}(\ell) + \underline{x}(\kappa)$, where $\underline{x}(\ell)$ is pointing to the ℓth cell of the crystal and $\underline{x}(\kappa)$ describes the relative position of the κth ion in the cell. The 3 × 3 two-ion force constant matrices $\underline{\phi}(L,L')$ are subject to the conservation laws of energy, momentum, and angular momentum and to the symmetry restrictions of the space group [2.2].

We note that our treatment is based on the assumption of crystal stability and on the adiabatic approximation (see [2.3]). Furthermore, the force constants being derived from a series expansion of the free energy $F(T,V)$ are pseudo-harmonic entities which contain a quasi-harmonic correction due to thermal expansion [2.4] as well as a self-energy renormalization arising from anharmonicity (see [2.1,5]).

From the vector equations of motion

$$M_\kappa \underline{\ddot{u}}(L) = -\frac{\partial \phi_2}{\partial \underline{u}(L)} = -\sum_{L'} \underline{\phi}(LL') \underline{u}(L') \tag{2.2}$$

using, e.g., periodic boundary conditions, one obtains the eigenvalues and eigenvectors of the system as the solutions $\omega_\lambda^2 \equiv \omega^2(\underline{q},j)$ and $\underline{e}_\lambda \equiv \underline{e}(\underline{q},j)$ of the secular equation

$$\det|\underline{D}(\underline{q}) - \underline{M}\omega^2 \underline{I}| = 0 \tag{2.3}$$

where the dynamical matrix $\underline{D}$ is defined as

$$\underline{D}(\kappa\kappa'|\underline{q}) = \sum_\ell \underline{\phi}(LL') \exp\{-i\underline{q}[\underline{x}(L) - \underline{x}(L')]\} . \tag{2.4}$$

$\underline{M} = (M_\kappa)$, i.e., it is a diagonal mass matrix, while $\underline{I}$ is a unit matrix, $\underline{I} = (\delta_{\alpha\beta}\delta_{\kappa\kappa'})$, where α,β are Cartesian indices. The index j labels the different branches belonging to every wave vector $\underline{q}$.

The most important entity to be known, in addition to the dispersion relation $\omega(\underline{q},j)$, is the phonon density of states

$$D(\omega) = \sum_j n_j(\omega) \tag{2.5}$$

with the contribution from branch j

$$n_j(\omega) = \sum_{\underline{q}} \delta[\omega - \omega(\underline{q}j)]$$

$$= V_a \int \frac{dS_{\underline{q}}}{|\nabla_{\underline{q}} \omega(\underline{q}j)|} , \tag{2.6}$$

where $dS_{\underline{q}}$ denotes a surface element in $\underline{q}$ space.

It should be noted that the frequencies do not uniquely determine the eigenvectors and force constants of a crystal [2.6]. Therefore, in a calculation of dynamical properties which contains phonon eigenvectors (such as Raman scattering) additional arguments from microscopic theory, etc., are required to demonstrate the physical significance of a set of (formal or model) force constants.

2.2 Rigid Ion Model (RIM)

The lattice energy of ionic crystals is mainly given by two-ion Coulomb interactions of pointlike charges. In diatomic cubic crystals with ionic charges, $Z_\kappa = \pm Ze$, this energy per lattice cell reads

$$U_{ion} = \frac{1}{2} \sum_{\ell,\kappa\kappa'}' \frac{Z_\kappa Z_{\kappa'} e^2}{|\underline{x}(\ell\kappa) - \underline{x}(0\kappa')|} = \alpha_M \frac{(Ze)^2}{r_o} \tag{2.7}$$

with the Madelung constant α_M and the nearest-neighbor distance r_o. At equilibrium, the resulting attractive forces are balanced by repulsive short-range electronic overlap forces. They are conveniently described by two-body central Born-Mayer potentials

$$V_{\kappa\kappa'}(r) = A_{\kappa\kappa'} \exp(-r/\rho_{\kappa\kappa'}) . \tag{2.8}$$

In this Born model the total lattice energy per cell is given by

$$U(r) = -\alpha_M \frac{(Ze)^2}{r} + \frac{1}{2} \sum_{\kappa\kappa'}' V_{\kappa\kappa'}(r) . \tag{2.9}$$

KELLERMANN used this model for a calculation of dispersion curves in alkali halides. Here, the force matrix can be split into a long-range Coulomb and a short-range repulsive part as

$$\underline{\phi} = \underline{\phi}^C + \underline{\phi}^R . \qquad (2.10a)$$

The Coulomb force matrix is given by

$$\underline{\phi}^C = \underline{Z}\underline{C}\underline{Z} , \quad \underline{Z} = (Z_\kappa) , \qquad (2.10b)$$

with the "Coulomb matrix" $\underline{C}$. The Fourier transform of $\underline{\phi}$ [refer to the corresponding transform in (2.4)] is the rigid-ion dynamical matrix

$$\underline{D}^{RI}(\underline{q}) = \underline{Z}\underline{C}(\underline{q})\underline{Z} + \underline{R}(\underline{q}) . \qquad (2.11)$$

$\underline{C}(\underline{q})$ is the Fourier transform of $\underline{C}$ and has a nonanalytical part at $q \to 0$ which leads to the splitting of the longitudinal-optic from the transverse-optic modes in the nonpolariton regime (Lyddane-Sachs-Teller relation). A specific technique originated by Ewald has been developed for the calculation of $\underline{C}(\underline{q})$ (refer to MARADUDIN et al. [2.7]). In the simplest nearest-neighbor approximation this "rigid ion" model reduces with the help of the equilibrium condition, $U'(r) = 0$, to a one-parameter model which gives a surprisingly good fit to the acoustic and the transverse optic modes. The calculated longitudinal optic mode frequency is, of course, too high since no electronic screening has been considered. In other words, $\varepsilon_\infty = 1$ in the RIM.

2.3 Dipole Models (DM)

The difficulties in describing properly the longitudinal optic modes and related entities may be overcome by an explicit consideration of the electronic charge distortion induced at a displaced ion (TOLPYGO, 1948; SZIGETI, 1948; see [2.8]). Several models have been developed during the last 20 years which treat this distortion in the dipole approximation (refer to [2.8-11]). The most pictorial of these models is the shell model.

2.3.1 Shell Model (SM)

This model, originally due to DICK and OVERHAUSER (1958) and COCHRAN (1959) (see [2.8]) introduces in addition to the ion vector displacement $\underline{u}(L)$ an electronic polarization coordinate $\underline{w}(L)$. The potential depends now in a bilinear form on all $\underline{w}(L)$ and $\underline{u}(L)$ of the lattice. The equations of motions are

$$M_\kappa \underline{\ddot{u}}(L) = - \frac{\partial \phi}{\partial \underline{u}(L)} \qquad (2.12)$$

$$M_{el}\ddot{\underline{w}}(L) = -\frac{\partial \phi}{\partial \underline{w}(L)} = 0 . \tag{2.13}$$

The second equation is the adiabatic condition for the shell model. One now introduces Coulomb and short-range forces between electrons as well as between electrons and ions and proceeds similarly to the case of the RIM; one obtains then the dynamical matrix

$$\underline{D}(\underline{q}) = \underline{D}^{RI}(\underline{q}) + \underline{D}^{DIP}(\underline{q}) , \tag{2.14}$$

where $\underline{D}^{RI}$ is the dynamical matrix of the RIM and $\underline{D}^{DIP}$ that of the induced dipolar forces

$$\underline{D}^{DIP} = -(\underline{T} + \underline{ZCY})(\underline{\bar{S}} + \underline{YCY})^{-1}(\underline{T}^{+} + \underline{YCZ}) . \tag{2.15}$$

The matrices $\underline{T}$ and $\underline{\bar{S}}$ represent the short-range electron-ion and electron-electron coupling. $\underline{Y}$ is an electronic charge matrix analogous to the ionic charge matrix $\underline{Z}$. The matrix $\underline{\bar{S}}$ consists of the short-range electron-electron coupling $\underline{S}$ and of the local shell-core coupling $\underline{K} = (K_{\kappa})$. In the simplest version of the shell model (SSM) one assumes that short-range forces act through the shells only ($\underline{R} = \underline{T} = \underline{S}$) and, furthermore, one neglects the the polarizability of the positive ions as compared to that of the negative ions ($K_{+} \rightarrow \infty$). In the nearest-neighbor approximation there are then two additional parameters, Y_{-} and K_{-}, as compared with Kellermann's RIM. They can be fitted to the optical constants in the following equations:

$$\mu\omega_{TO}^{2} = R_{o}' - \frac{4\pi}{3} \frac{\varepsilon_{\infty} + 2}{3} Z_{s}^{2} \frac{e^{2}}{V_{a}} \tag{2.16}$$

$$\mu\omega_{LO}^{2} = R_{o}' + \frac{8\pi}{3} \frac{\varepsilon_{\infty} + 2}{3} Z_{s}^{2} \frac{e^{2}}{V_{a}} . \tag{2.17}$$

Here μ is the reduced mass of the lattice cell. The effective Szigeti charge Z_{s} is given by

$$Z_{s} = Z_{+} + \frac{Y_{-}}{1 + K_{-}/R_{o}} . \tag{2.18}$$

R_{o}' defines the "center" frequency

$$\omega_{c}^{2} \equiv \frac{1}{3}(2\omega_{TO}^{2} + \omega_{LO}^{2}) \tag{2.19}$$

$$\mu\omega_{c}^{2} = R_{o}' = \frac{R_{o}}{1 + R_{o}/K_{-}} . \tag{2.20}$$

This three-parameter SM (A, Y_-, K_-) improves the agreement between experiment and theory remarkably as compared with the RIM (see, e.g., Fig.2.10 for NaI). A is the longitudinal force constant to nearest neighbors.

2.3.2 Extended Shell Model (ESM)

An alternative representation of the dynamical matrix, (2.14), has been suggested by COCHRAN [2.8]

$$D(\underline{q}) = \underline{R}' + \underline{Z}'\underline{C}'\underline{Z}' \tag{2.21}$$

with the effective entities

$$\underline{R}' = \underline{R} - \underline{T}\underline{\bar{S}}^{-1}\underline{T}^+ \tag{2.22}$$

$$\underline{Z}' = \underline{Z} - \underline{T}\underline{\bar{S}}^{-1}\underline{Y} \tag{2.23}$$

$$\underline{C}' = (\underline{I} + \underline{C}\underline{Y}\underline{\bar{S}}^{-1}\underline{Y})^{-1}\underline{C} \ . \tag{2.24}$$

The general structure of (2.21) is completely analogous to that of the microscopic theory (refer to [2.11,12]).

This ESM is, of course, able to describe every type of phonon dispersion relation by introducing a sufficient number of parameters in the matrices $\underline{R}, \underline{\bar{S}}, \underline{T}$. A satisfactory description of the data requires, however, the introduction of pseudo-polarizabilities with positive electronic shell charges Y_+ at the cation lattice sites [2.13].

2.3.3 Overlap Shell Model (OSM)

A real justification of this argument can be obtained by a discussion of the polarization effects in the overlap charge [2.14]. A particularly fruitful version of this type of shell model turned out to be its combination with valence force fields for the case of partially ionic tetrahedrally coordinated compounds [2.15] (refer to Sect. 2.3.9). A particularly appealing feature of the OSM is the drastic reduction of second nearest-neighbor forces which often allows a good description of data with only six parameters (A, B, Y_+, Y_-, K_+, K_-). A and B are nearest neighbor longitudinal and transversal coupling constants respectively.

2.3.4 Deformation Dipole Model (DDM)

The shell model, in its extended version, is equivalent to the deformation dipole model (DDM) originally developed by HARDY [2.9]. The equivalence can be shown if the DDM is completed with a polarization self-energy which is missing in the original version [2.16]. The inter-relation between both models has been illuminated in a recent microscopic calculation by ZEYHER [2.17].

2.3.5 Deformable Shell Model (DEFSM)

The concept of a dipolar electronic charge distortion in the dipole models may be generalized to models with general electronic charge distortions consistent with the local symmetry character of a phonon with respect to a particular "deformable" ion. For example, in a cubic diatomic lattice the radial coupling of an ion to its six nearest neighbors may be decomposed into three dipolar, one monopolar (breathing), and two quadrupolar deformabilities. We may define a generalized short-range dynamical matrix $\underline{R}'$ which in the case of cubic ionic crystals has the form

$$\underline{R}' = \underline{D}^{RI} + \underline{D}^{DEF} \tag{2.25}$$

with

$$\underline{D}^{DEF} \simeq -\sum_{\Gamma} \underline{T}_{\Gamma} \underline{\bar{S}}_{\Gamma}^{-1} \underline{T}_{\Gamma}^{+} \tag{2.26}$$

where Γ denotes the different symmetry representations [i.e., the monopolar (Γ_1^+), dipolar (Γ_{15}^+), and quadrupolar (Γ_{12}^+) symmetry in the radial nearest-neighbor approximation] of the ionic deformabilities. The concept becomes useful in cases where a specific deformability resembles an important part of the excitation spectrum of an ion. The first attempt in the case of alkali halides was the introduction of the breathing shell model.

2.3.6 Breathing Shell Model (BSM) and Double Shell Model (DSM)

The Γ_1^+ deformability of the anions was introduced by SCHRÖDER [2.18] in order to improve the description of the longitudinal optic branches in particular near the L point in the Brillouin zone. While it seems that in these cases the introduction of dipolar pseudo-polarizabilities at the cation lattice sites may be preferable [2.14] (see Sect. 2.3.3), the idea of a "breathing" deformability may be appropriate in systems with a very localized dynamic charge transfer such as $Sm_xY_{1-x}S$. Furthermore, in superconducting crystals like Nb [2.19] and NbC [2.20], a combination of volume and quadrupole deformabilities (in the second case at the nonmetal lattice sites) seems to give a very satisfactory approach to the pronounced anomalies in the dispersion curves of these crystals.

An alternative approach in this case has been given by WEBER in his double-shell model (DSM) [2.21] which has been used with great success during the last few years. We note that many modifications of the BSM have been published under different names ("deformable shell model" DEFSM, etc.) (refer to BASU et al. [2.22]), but no new basic idea seems to originate with these proposals.

2.3.7 Quadrupole Shell Model (QSM)

A very interesting application of the quadrupolar deformability Γ_{12}^+ has been found in the case of silver halides by FISCHER et al. [2.23]. Here, the local virtual excitation of a d electron into an excited s state leads to a deformability of the right symmetry. Nearest-neighbor d-p hybridization may also contribute. The prediction by FISCHER et al. of an inversion of the transverse-mode eigenvectors in AgBr near the L point has been recently confirmed [2.24]. A microscopic calculation starting from the electronic band structure of silver halides has shown that the model is essentially correct [2.25]. Furthermore, a quadrupolar deformability may play a role in the lattice dynamics of copper halides and in the vibronic spectra of some oxides [2.26]. On the other hand, it seems doubtful whether the introduction of quadrupolar forces into the lattice dynamics of alkali halides is a useful concept [2.27].

2.3.8 Bond Charge Model (BCM)

In covalent crystals like diamond and its homologues the shell model and its modifications are useful models [2.28,29], although the electronic charge accumulation between the ions suggests a different description, i.e., the bond charge model (BCM). In its first version the bond charges (represented by point charges) had their positions always at the midpoint between neighboring ions [2.10]. In a recent extension by WEBER, the bond charge takes up the proper adiabatic force-free position in all displaced configurations. The two short-range forces are those of the KEATING model [2.31] which are essentially valence forces (see Sect. 2.3.9). A four-parameter version of the BCM gives a very satisfactory description of the dispersion curves of diamond, silicon, and germanium. The model, while sufficient for describing dispersion curves, fails in reproducing the complete crystal polarizability. Possibly the introduction of a charge transfer or a charge form factor could remove this failure.

2.3.9 Valence Force Field Model (VFFM)

This approach can be regarded as an extrapolation of the theory of normal vibrations of molecules. Here, the potential energy of the molecule is often given in terms of internal displacement coordinates, such as changes of bond distances, bond angles, etc. (refer to WILSON et al. [2.32]). The interatomic forces stem from distortions of the electronic charge density which is strongly concentrated along the inter-ionic connecting lines due to hybridization of the atomic orbitals. Since the covalent crystals behave in many respects like very large molecules, one should be able to describe them by an appropriate extension of molecular methods to almost infinite systems. The valence force field constants in these crystals may be obtained from an analysis of organic molecules with similar covalent bonds. For example, SCHACHTSCHNEIDER and SNYDER [2.33] have made a least-square determination of valence forces in branched paraffins which has been used successfully to determine the phonons in diamond, silicon, germanium, and gray tin [2.34].

It should be noted that the valence force field potential does not contain any Coulomb interactions which means that the interacting electronic and ionic charges are well screened. On the other hand, the VFFM lacks a clear relation to the dielectric properties of the crystals which gives the shell models their strong appeal.

2.3.10 Valence Overlap Shell Model (VOSM)

In the case of mixed ionic-covalent crystals a combination of the DDM with the VFFM has been used by KUNC et al. [2.35] for the zinc-blende structure compounds like ZnSe with fair success. A combination of the OSM with the VFFM (VOSM) describes quite successfully the lattice vibrations of many III-V, II-VI, and I-VII crystals [2.15]. In addition, this combination provides a useful basis for the analysis of second-order Raman spectra [2.15].

2.3.11 Three-Body Force Shell Model (TSM)

The deformabilities described in Sects. 2.3.4-2.3.7 are different combinations of three- and four-body forces. For example, the "breathing" deformability leads to a decrease of the bulk elasticity, $c_{11} + 2c_{12}$, while the quadrupolar (Γ_{12}^+) deformability decreases the shear elastic constant, $c_{11} - c_{12}$. Another possibility to introduce many-body forces is to start directly from the violation of the Cauchy relation, $c_{12} \neq c_{44}$ in cubic crystals, to determine many- (mainly three-) body forces (refer to [2.1,22]). This leads automatically to a type of deformable shell model but without a clear description in terms of local symmetry coordinates. SINGH and VERMA [2.36] introduced the so-called three-body-force shell model (TSM) which is, however, lacking a microscopic justification if $c_{12} - c_{44} < 0$ (refer to ZEYHER [2.17]). In this respect it is similar to the BSM if used for some alkali halides, but one should note that anharmonic corrections, which are very important for the Cauchy "violation", are not considered in those models. For recent refinements of the TSM see SINGH et al. in [2.38].

2.4 Microscopic Theory and Models

In this section a short comment is given on the present state of the art of the microscopic theory of phonons. For recent reviews the reader is referred to articles by SHAM and by BILZ et al. in [2.39] and to [2.1,10,12].

In microscopic theories one begins with the Coulomb electron-electron and electron-ion interactions and eventually derives the phonon dispersion relations by some approximation to the many-body Hamiltonian. The essential assumption is the definition of "ions" which include the tightly bound electrons. The Coulomb potential of the ions has then to be replaced by a pseudopotential which takes account of the exchange and correlation effects of the valence electrons with the core electrons.

At the present time, two different approaches are used in microscopic theory. The first one describes the short-range inter-atomic forces as originating with the overlap of (mainly) nearest and (to a smaller extent) second-nearest neighbor charge densities. The overlap effect also leads to a modification of long-range Coulomb forces. A successful sequence of investigations [2.17] has recently led to the first parameter-free calculation of phonon dispersion curves in LiD [2.40].

The second approach is the dielectric function method. The basic ideas of this method have been developed in the 1960s (refer to SINHA [2.10]). During the last years the theory has concentrated on localized wave functions in order to obtain an explicit representation of the inverse dielectric function which appears in the dynamical matrix (refer to HANKE [2.12]). These treatments provide a microscopic basis for a generalized dipole model which may be applied to metals as well as to insulators and semiconductors.

2.5 Abbreviations Used in Figure Captions

$\omega(\underline{q})$	Phonon frequency as a function of wave vector
$\omega_c(\underline{q})$	Calculated phonon frequency only
$D(\omega)$	Density of states as a function of frequency
$D_m(\omega)$	Measured density of states (amplitude weighted)
[Ref. 7.8, Fig. 4]	Figure 4 in [7.8] at the end of this book
T = 300 K	Measured at 300 K temperature
P: 1.36 Kb	Measured at 1.36 Kb pressure
M: 7 P-FCM	Model used: seven parameter force constant model
3NN-MASM	3rd nearest neighbour modified axially symmetric force constant model
-FCM	Force constant model
-RIM	Rigid ion model
-SM	Shell model
-MSM	Modified shell model
-BSM	Breathing shell model
-DDM	Deformation dipole model
-TSM	Three-body force shell model
-DEFSM	Deformable shell model
-DBM	Deformable bond model
-BCM	Bond charge model
-VSM	Valence force shell model
-VDDM	Valence force deformation dipole model
-DSM	Double shell model
-ASM	Axially symmetric model
-MASM	Modified axially symmetric model
-VFFM	Valence force model
-OSM	Overlap shell model
-VOSM	Valence overlap shell model
-RLM	Rigid layer model
-QSM	Quadrupole shell model
M:-	No model used (guide lines to the eye)
Lit. [10.9]	For further information about this crystal see [10.9]

2.6 Conversion Factors Between Units Used to Express Phonon Energies

	1 THz	10^{13} rad s^{-1}	10^2 cm^{-1}	10 meV
1 THz	1.0000	0.6283	0.3336	0.4136
10^{13} rad s^{-1}	1.5916	1.0000	0.5309	0.6582
10^2 cm^{-1}	2.9979	1.8836	1.0000	1.2398
10 meV	2.4181	1.5193	0.8066	1.0000

Part II

Phonon Atlas of Dispersion Curves and Densities of States

3. Rare-Gas Crystals

Low-temperature measurements of all rare-gas crystals exist. The main interest is clearly in the different modifications of solid He. Here, the theory of quantum crystals has to be used [3.1,2], while the other rare-gas crystals are usually analyzed in terms of (modified) Lennard-Jones (6,12) potentials corresponding to 3NN-FCM.

The following systems are treated:

Crystal	Figures showing	
	Dispersion curve $\omega(\underline{q})$	Density of states $D(\omega)$
He^4 (fcc)	3.1	
He^4 (hcp)	3.2a	3.2b
He^4 (bcc)	3.3	
Ne	3.4a	3.4b
Ar	3.5a	3.5b
Kr	3.6a	3.6b
Xe	3.7a	3.7b

He⁴(fcc)

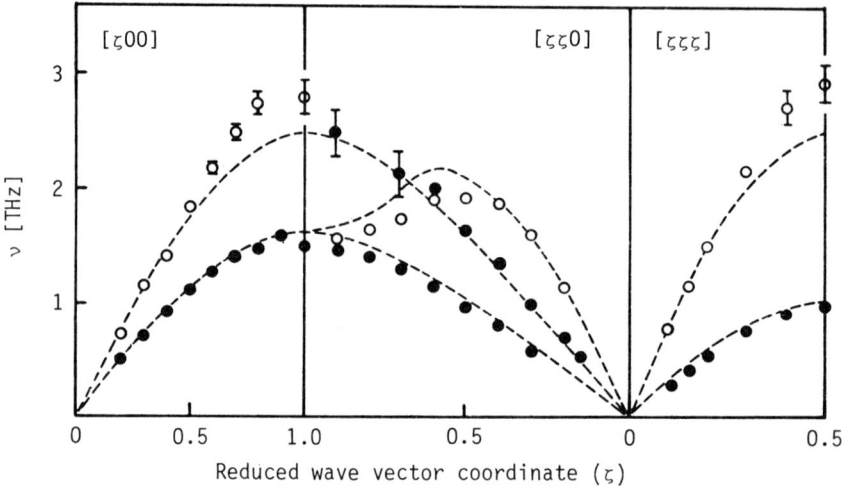

Fig. 3.1. He^4(fcc): $\omega(\underline{q})$ [Ref. 3.1, Fig. 4], T = 16 K, V_m = 11,7 cm^3, dashed lines calc. [3.2]

He⁴(hcp)

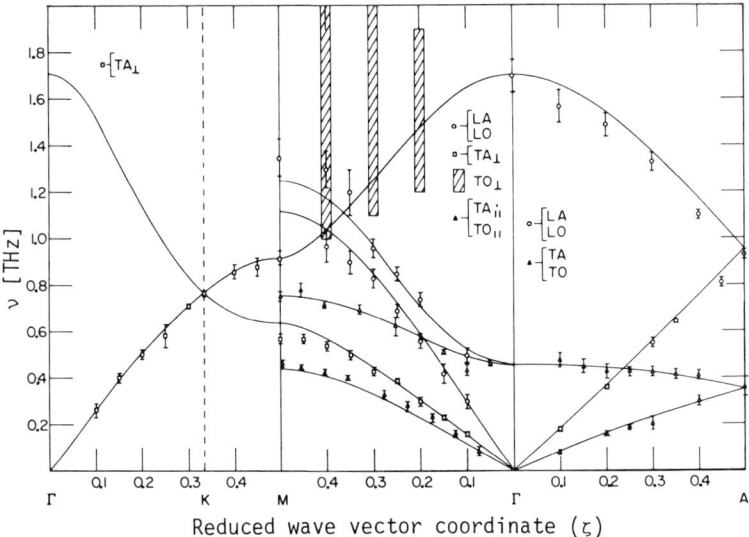

Fig. 3.2a. He^4(hcp): $\omega(\underline{q})$ [Ref. 3.3, Fig. 4], T = 4.2 K, p = 0.23 kb, M: 18P-MASM, TA$_{\parallel}$, TO$_{\parallel}$ from Ref. 3.4

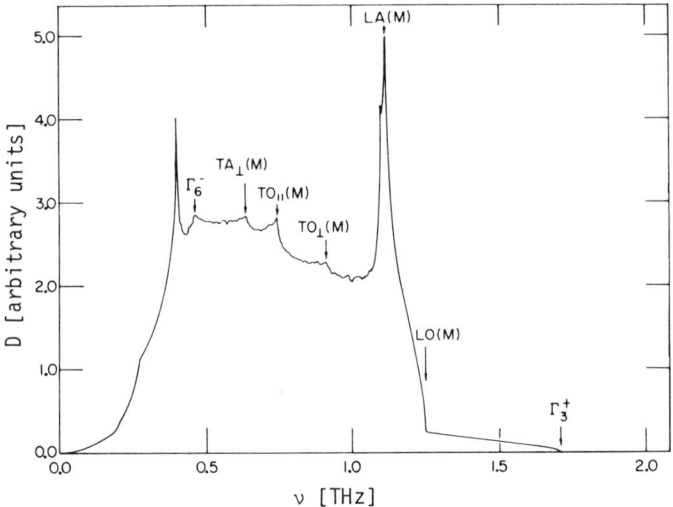

Fig. 3.2b. He^4(hcp): $D(\omega)$ [Ref. 3.3, Fig. 6], T = 4.2 K, p = 0.23 kb, M: 18P-MASM

He⁴(bcc)

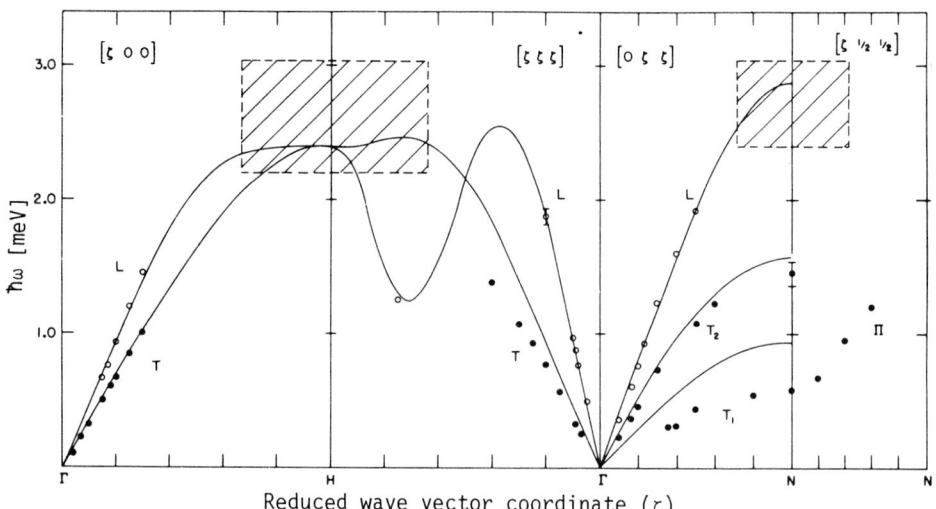

Fig. 3.3. He^4(bcc): $\omega(q)$ [Ref. 3.5, Fig. 7], T = 1.63 K, a = 4.12 Å, solid lines calc. [3.6,7]

Ne

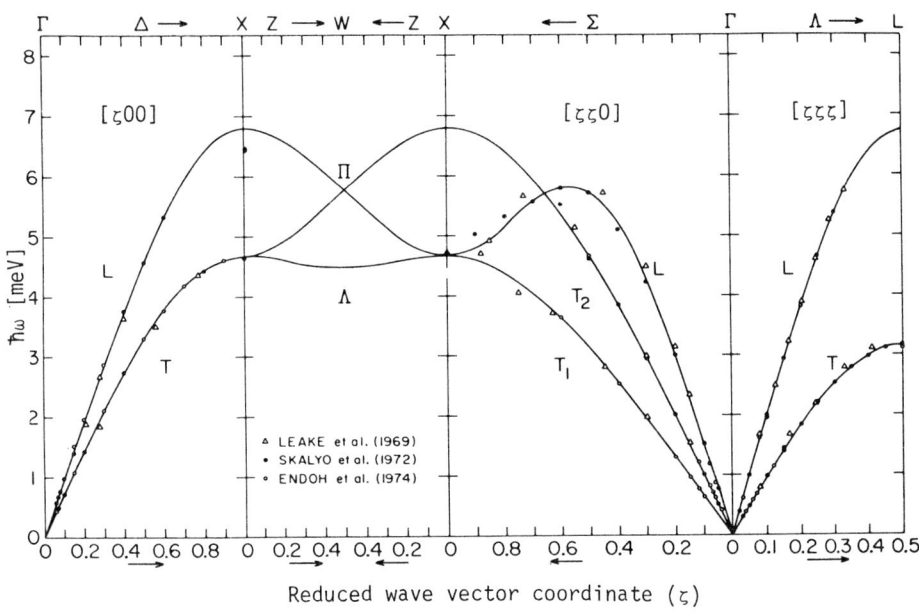

Fig. 3.4a. Ne: $\omega(\underline{q})$ [Ref. 3.11, Fig. 7], T = 5 K, M: 3NN-FCM, Lit. [3.8,9,10]

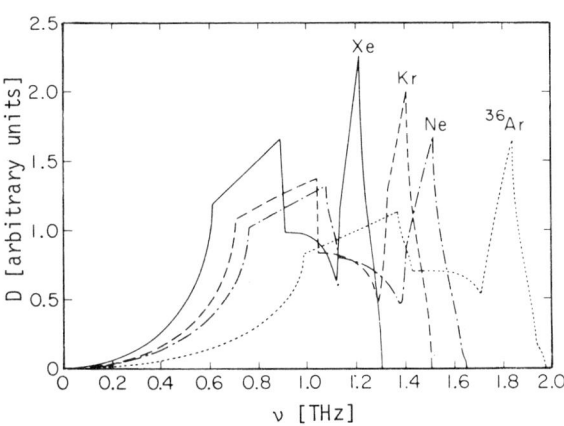

Fig. 3.4b. Ne: $D(\omega)$ [Ref. 3.11, Fig. 8], T = 5 K, M: 3NN-FCM, Lit. [3.8,9,10]

Ar

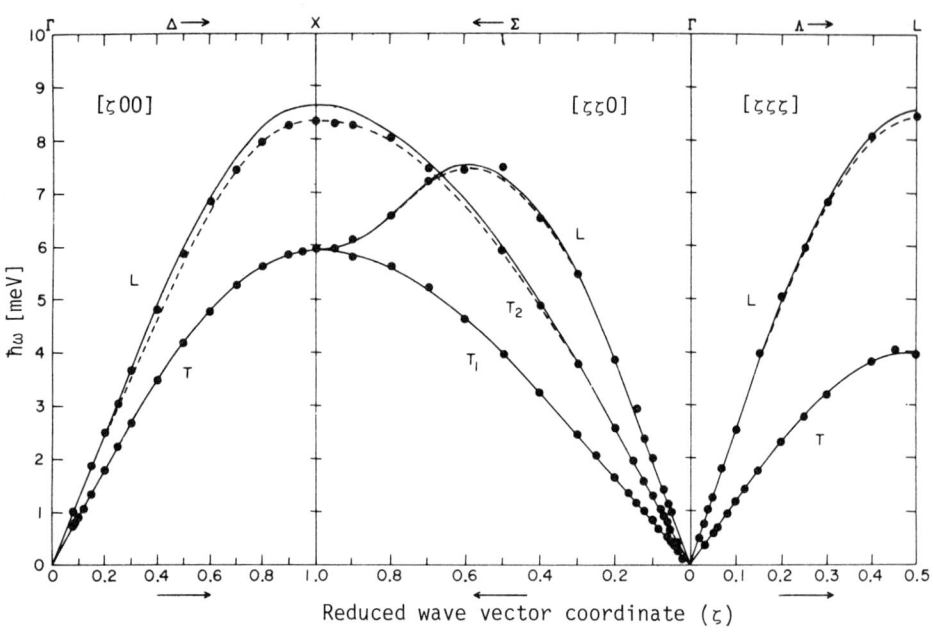

Fig. 3.5a. Ar: ω(q) [Ref. 3.12, Fig. 2], T = 10 K, M: 3NN-FCM (dashed lines), solid lines calc. [13]

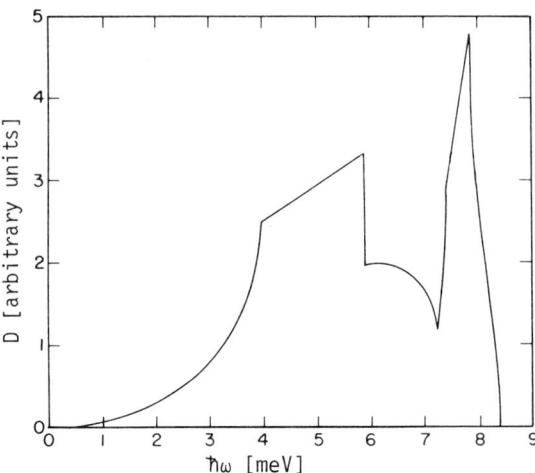

Fig. 3.5b. Ar: D(ω) [Ref. 3.12, Fig. 3], T = 10 K, M: 3NN-FCM

Kr

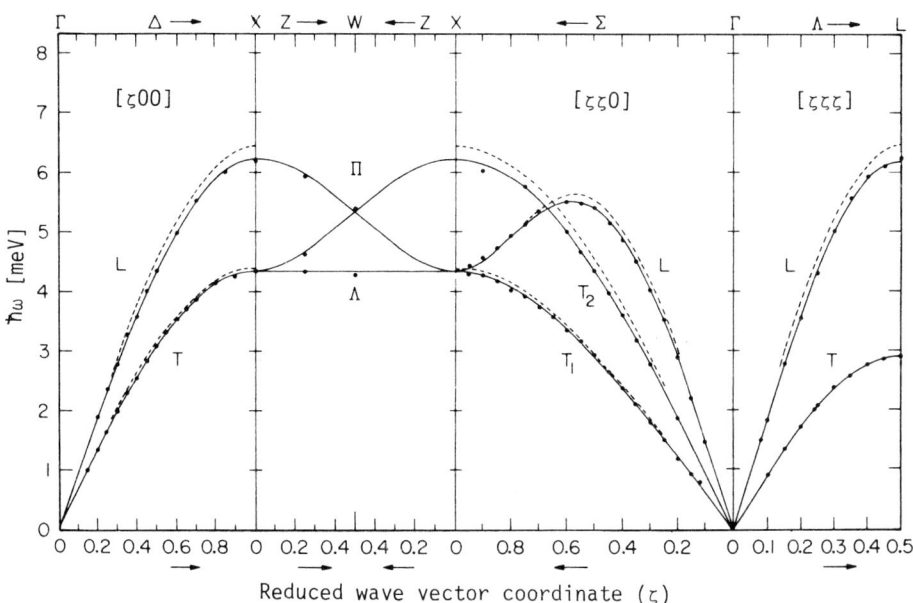

Fig. 3.6a. Kr: ω(**q**) [Ref. 3.14, Fig. 3], T = 10 K, M: 3NN-FCM (solid line), dashed line calc. [15]

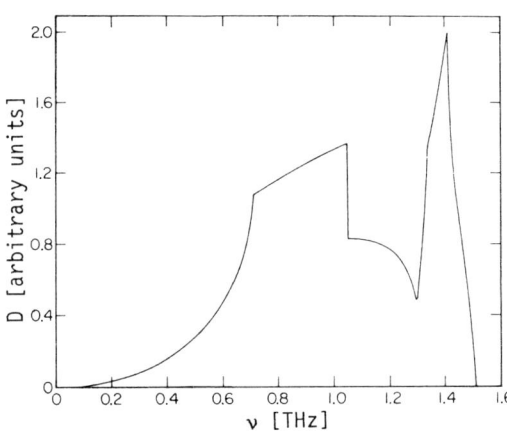

Fig. 3.6b. Kr: D(ω) [Ref. 3.14, Fig. 4], T = 10 K, M: 3NN-FCM

Xe

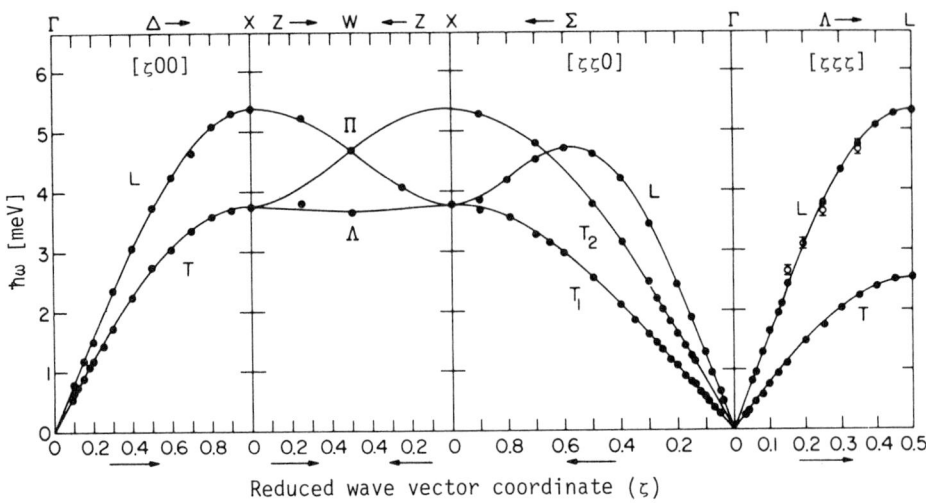

Fig. 3.7a. X: $\omega(\underline{q})$ [Ref. 3.16, Fig. 1], T = 10 K, M: 3NN-FCM, open circles [3.17], T = 4 K

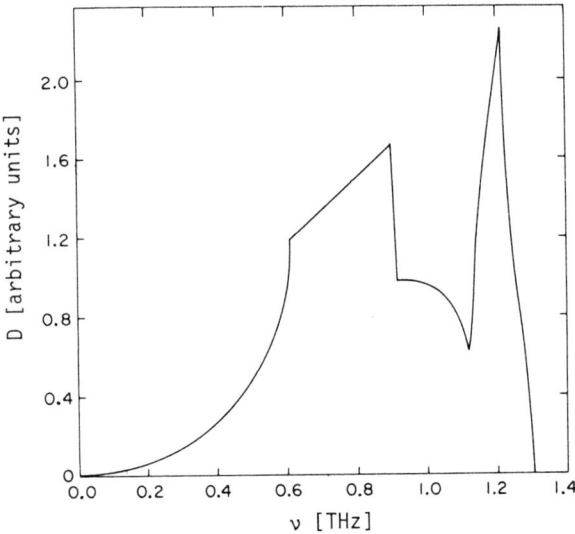

Fig. 3.7b. Xe: $D(\omega)$ [Ref. 3.16, Fig. 2], T = 10 K, M: 3NN-FCM

4. Alkali Halides (Rock Salt Structure)

Alkali halides are the best investigated family of crystals. Experiments are usually done at room temperature or at that of liquid nitrogen. Since the model theory (mainly dipolar models such as the SM or DM) as well as the microscopic theory [4.4] are highly developed (for recent reviews see [2.41, 4.8]) it seems to be worthwhile to measure phonons at helium temperature in a few alkali halides such as LiD, NaI, etc. This would provide a basis for a quantitative analysis of the anharmonic properties of these crystals which is still in its infancy.

The following systems are treated:

Crystal	Figures showing Dispersion curve $\omega(\underline{q})$	Density of states $D(\omega)$	Crystal	Figures showing Dispersion curve $\omega(\underline{q})$	Density of states $D(\omega)$
^{7}LiH	4.1a	4.1c	NaI	4.10a	4.10b
^{7}LiD	4.1b	4.1c	KF	4.11a	4.11b
LiF	4.2a	4.2b	KCl	4.12a	4.12b
LiCl	4.3		KBr	4.13a,b	4.13c
LiBr	4.4		KI	4.14a	4.14b
LiI	4.5		RbF	4.15a	4.15b
NaH		4.6	RbCl	4.16a	4.16b
NaF	4.7a	4.7b	RbBr	4.17a	4.17b
NaCl	4.8a	4.8b	RbI	4.18a	4.18b
NaBr	4.9a	4.9b			

^{7}LiH ^{7}LiD

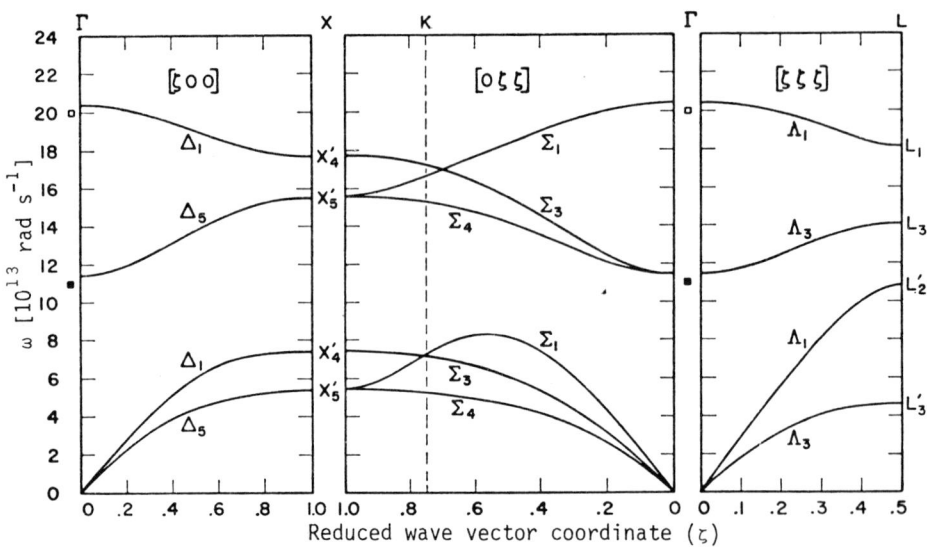

Fig. 4.1a. ^{7}LiH: $\omega_c(\underline{q})$ [Ref. 4.1, Fig. 4], T = 300 K, M: 7P-SM, Lit. [4.2]

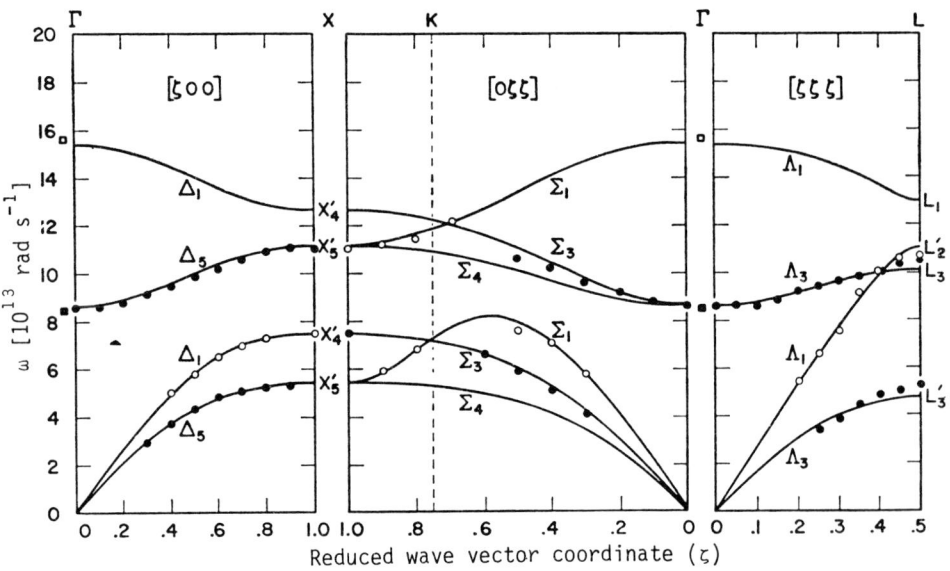

Fig. 4.1b. ^{7}LiD: $\omega(\underline{q})$ [Ref. 4.1, Fig. 3], T = 300 K, M: 7P-SM, Lit. [4.3-8]

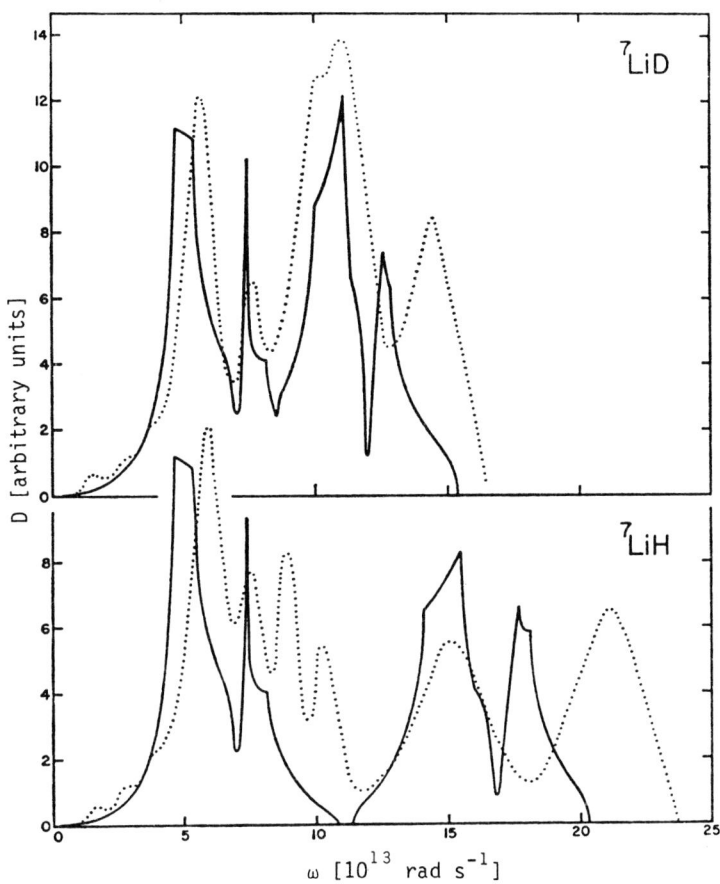

Fig. 4.1c. ^{7}LiH ^{7}LiD: $D(\omega)$ [Ref. 4.1, Fig. 5], dashed lines: measured [4.9], solid lines: calculated curve, T = 300 K, M: 7P-SM

LiF

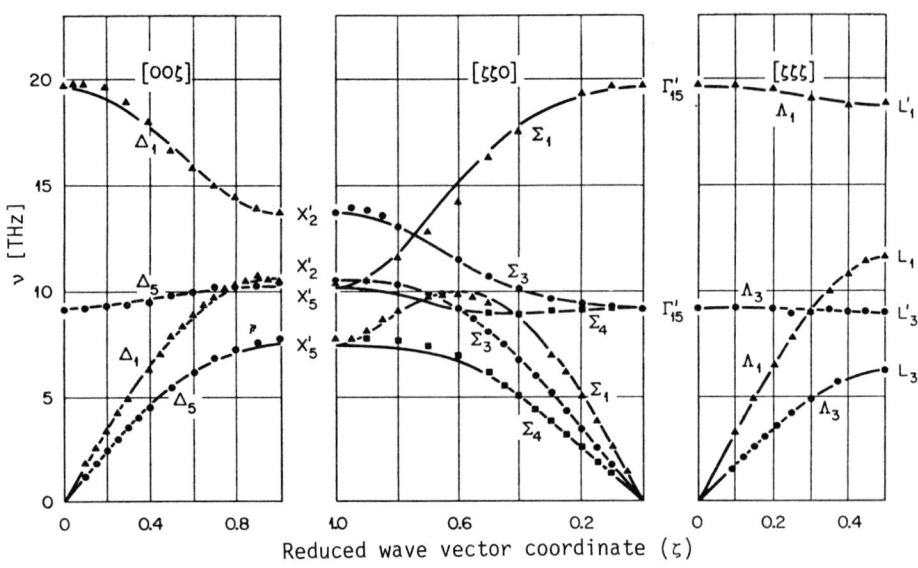

Fig. 4.2a. LiF: ω(q) [Ref.4.10, Fig.4], T = 298 K, M: 7P-SM, Lit. [4.2,11-13,64]

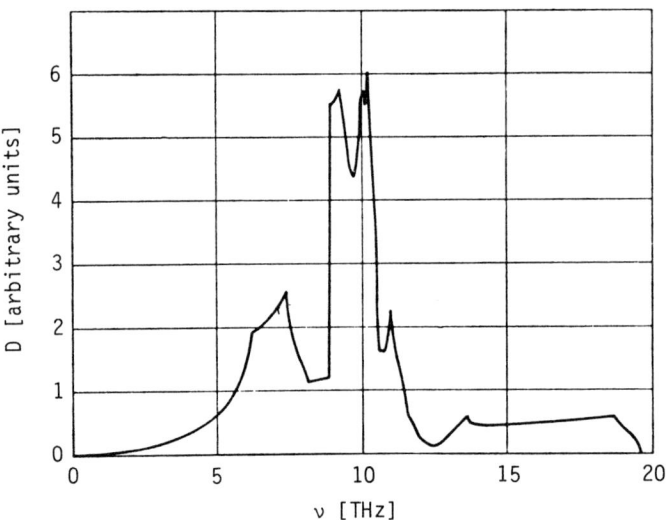

Fig. 4.2b. LiF: D(ω) [Ref.4.10, Fig.6a], T = 298 K, M: 7P-SM, Lit.[4.2]

LiCl

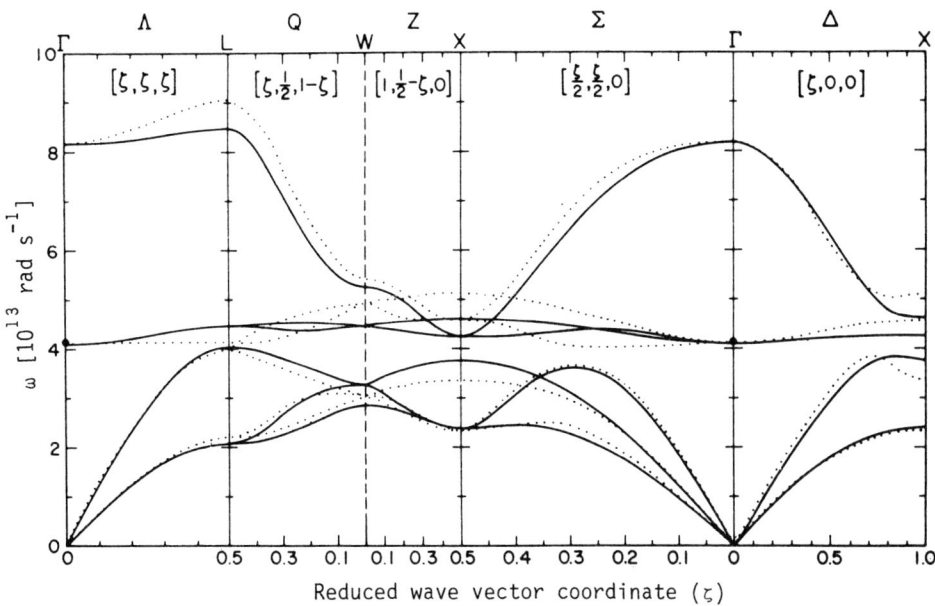

Fig. 4.3. LiCl: $\omega_c(\underline{q})$ [Ref. 4.13, Fig. 1], T = 0 K, M: 7P-BSM (full lines), 7P-DDM (dotted lines), Lit. [4.2]

LiBr

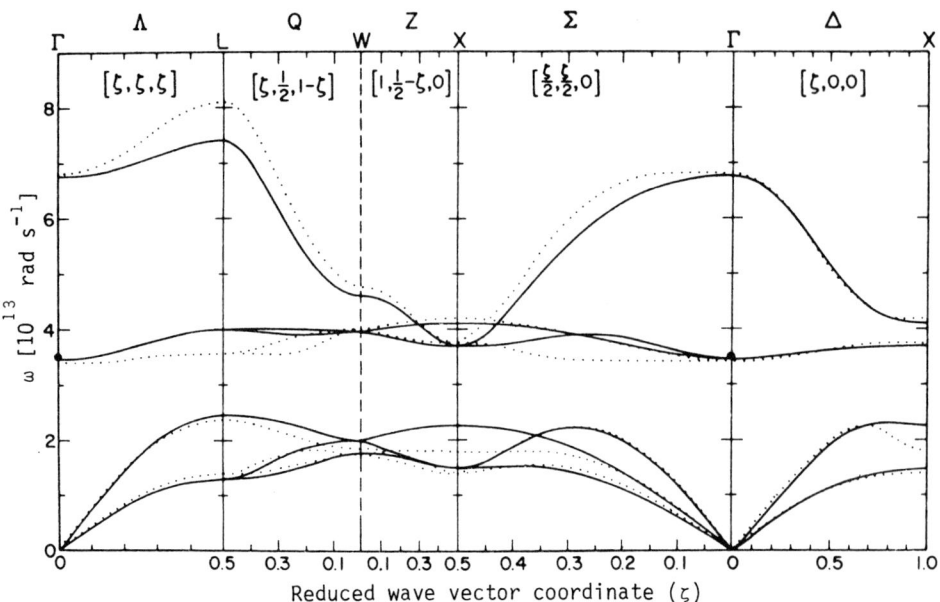

Fig. 4.4. LiBr: $\omega_c(\underline{q})$ [Ref. 4.13, Fig. 1], T = 0 K, M: 7P-BSM (full lines), 7P-DDM (dotted lines), Lit. [4.14]

LiI

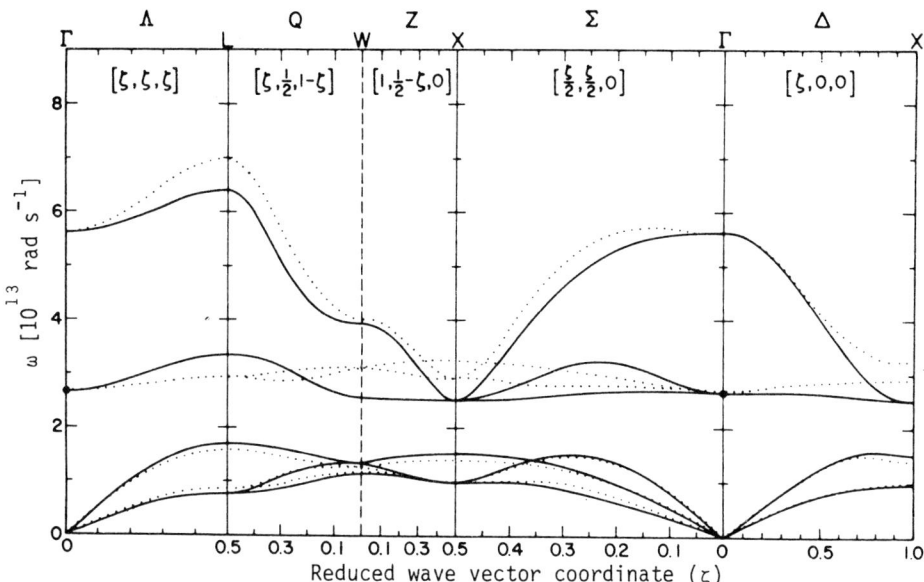

Fig. 4.5. LiI: $\omega_c(\underline{q})$ [Ref. 4.13, Fig. 1], T = 290 K, M: 7P-BSM (full lines), 7P-DDM (dotted lines), Lit. [4.14]

NaH

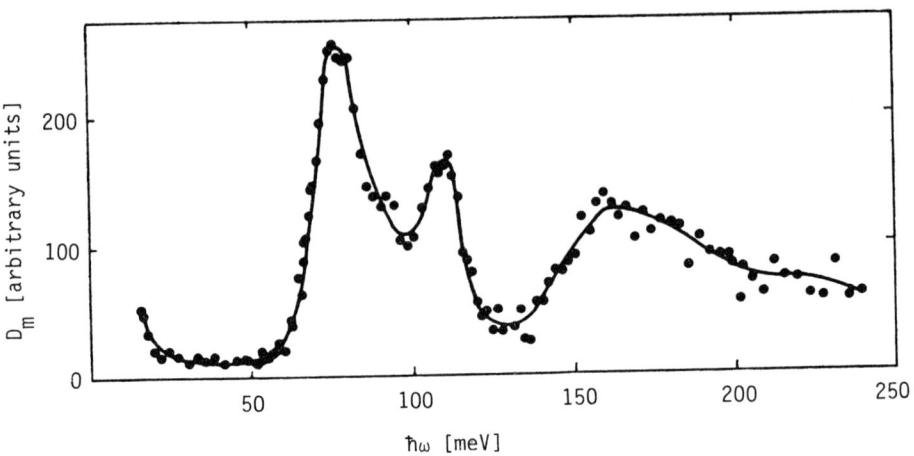

Fig. 4.6. NaH: $D(\omega)$ [Ref. 4.15, Fig. 7], T = 90 K

NaF

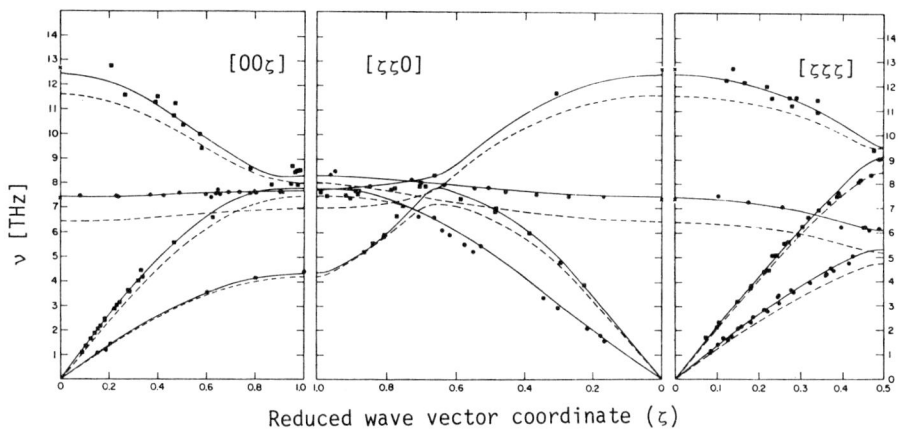

Fig. 4.7a. NaF: ω(q) [Ref.4.16, Fig.5], T = 295, M: 9P-SM (full lines), 5P-SM (dashed lines), Lit. [4.17-24]

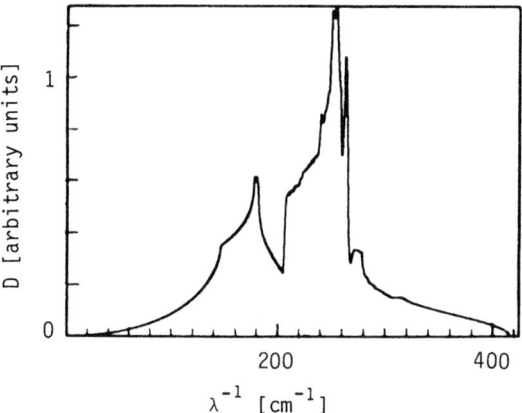

Fig. 4.7b. NaF: D(ω) [Ref. 4.25, Fig 4], M: 11P-SM, Lit.[4.19,20]

NaCl

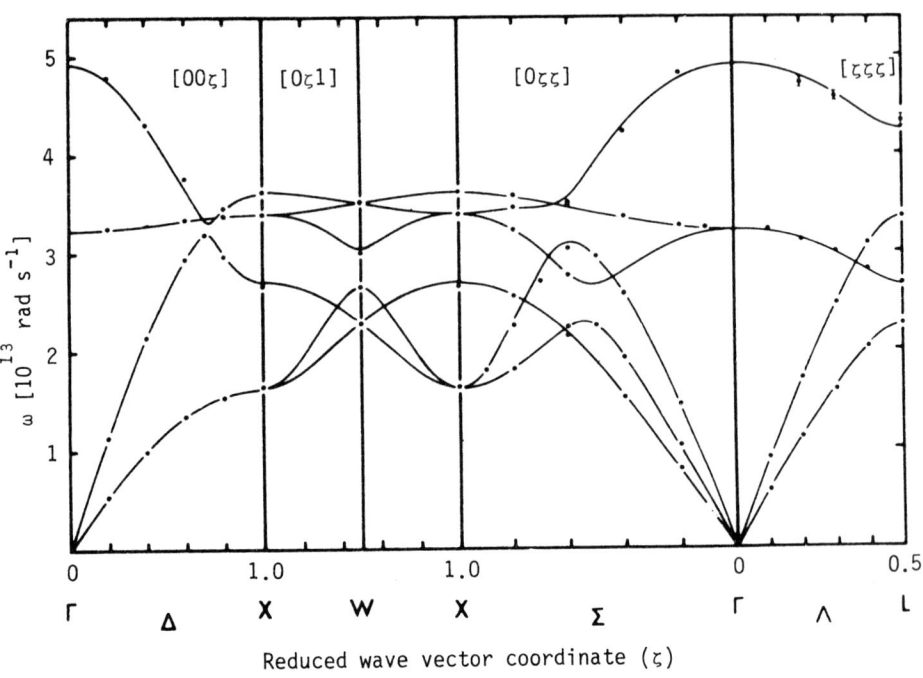

Fig. 4.8a. NaCl: $\omega(\underline{q})$ [Ref. 4.26, Fig. 1a], T = 80 K, M: 11P-SM, Lit. [4.11,18,20,21,24,25,27-33]

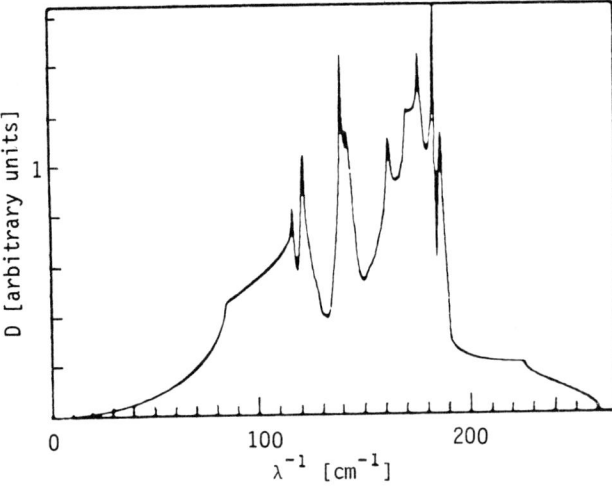

Fig. 4.8b. NaCl: $D(\omega)$ [Ref. 4.25, Fig. 4], T = 80 K, M: 11P-SM, Lit. [4.26]

NaBr

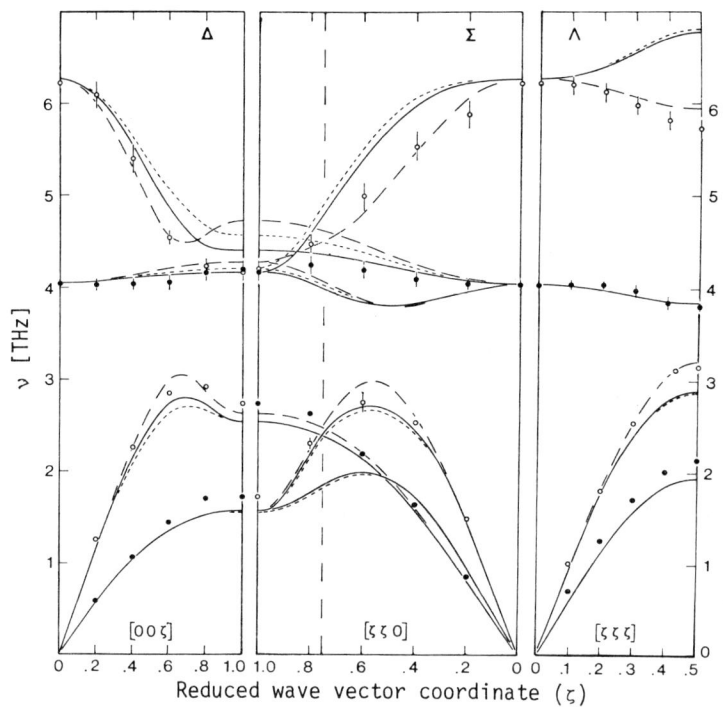

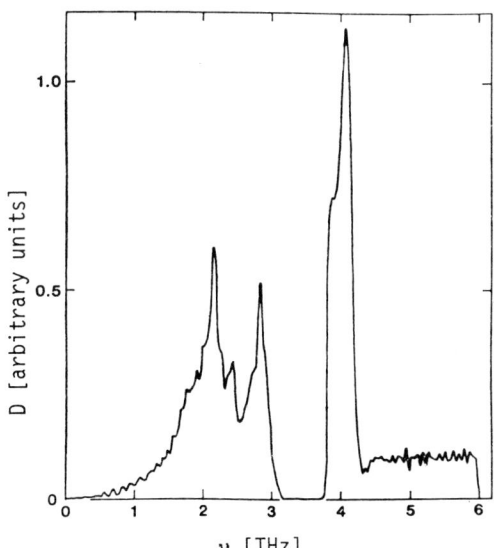

Fig. 4.9a. NaBr: ω(q) [Ref. 4.34, Fig. 7], T = 295 K, M: 7P-SM (solid lines), 9P-SM (dashed lines), 8P-BSM (dotted lines), Lit. [4.18,20,21,24,25,35,63]

Fig. 4.9b. NaBr: D(ω) [Ref. 4.34, Fig. 8], M: 9P-SM

NaI

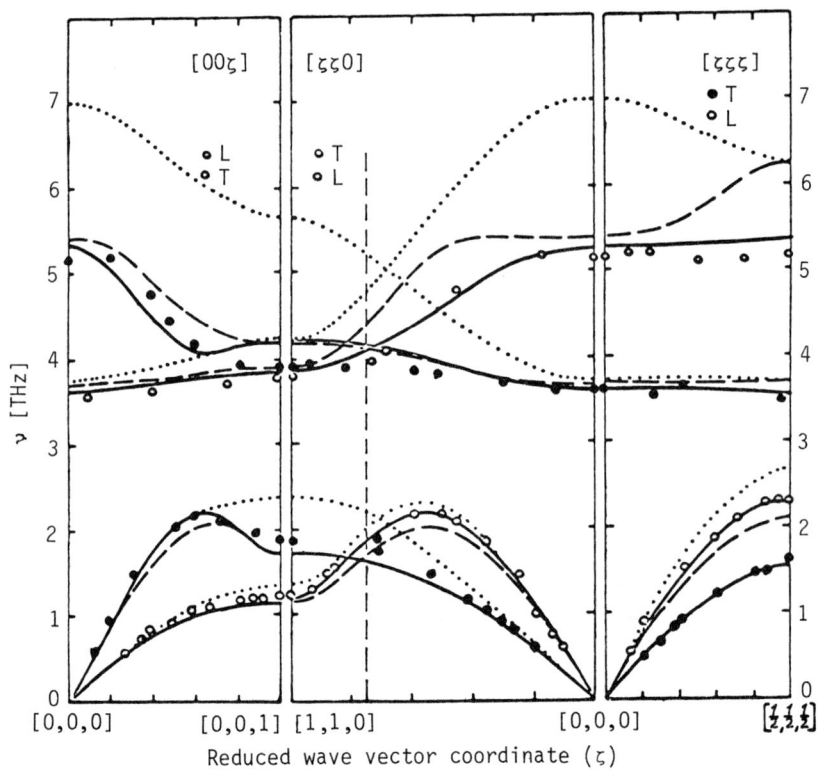

Fig. 4.10a. NaI: ω(q) [Ref. 4.36, Fig. 1], T = 100 K, Measurements [4.37,38], M: 7P-BSM (full lines), 7P-SM (dashed lines), 5P-FCM (dotted lines), Lit. [4.18,20,24,36-43]

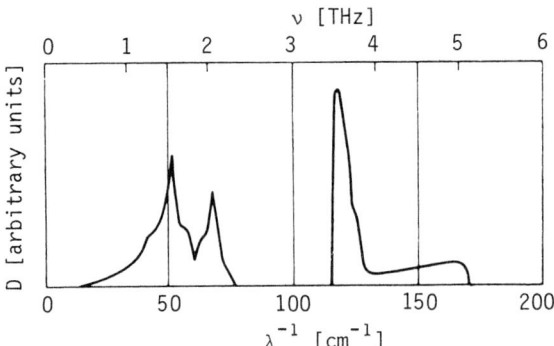

Fig. 4.10b. NaI: D(ω) [Ref. 4.44, Fig. 4b], M: 9P-SM

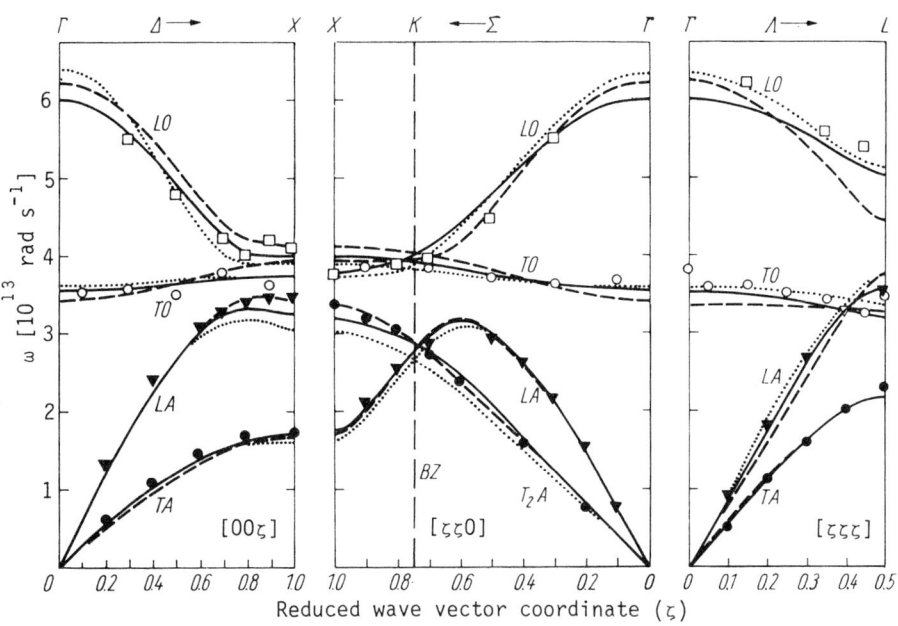

Fig. 4.11a. KF: ω(q) [Ref. 4.45, Fig. 1], T = 300 K, M: 11P-SM (solid-lines), 7P-FCM (dashed lines), 9P-TSM (dotted lines), Lit. [4.46]

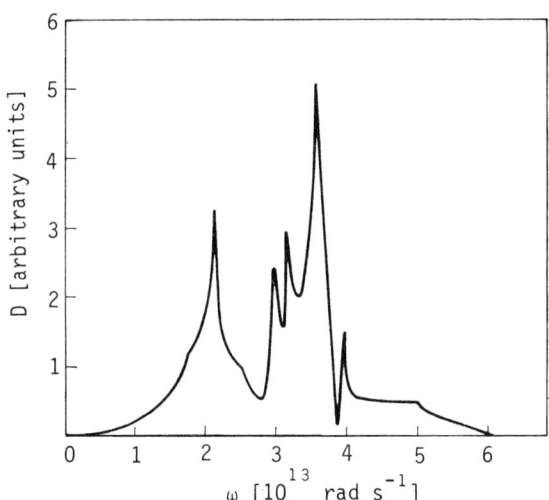

Fig. 4.11b. KF: D(ω) [Ref. 4.45, Fig. 2], M: 11P-SM

KCl

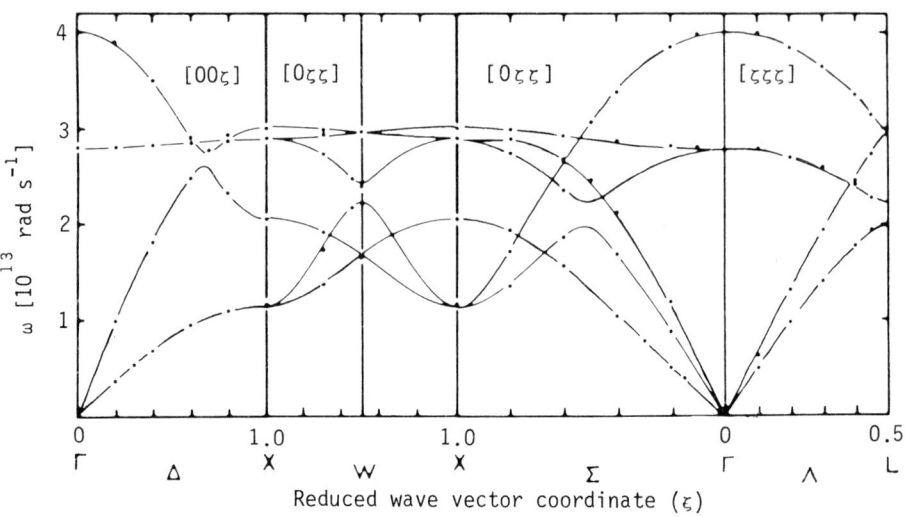

Fig. 4.12a. KCl: ω(q) [Ref. 4.26, Fig. 1b], T = 80 K, M: 11P-SM, Lit. [4.30, 46-50,64]

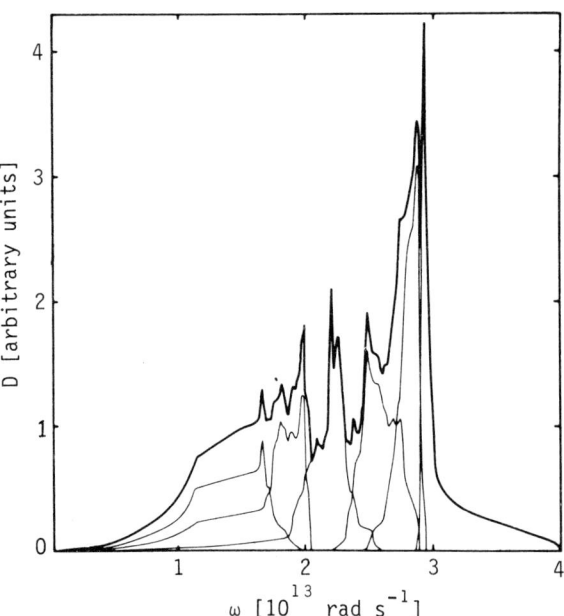

Fig. 4.12b. KCl: D(ω) [Ref. 4.26, Fig. 2b], T = 80 K, M: 11P-SM, Lit. [4.25]

KBr

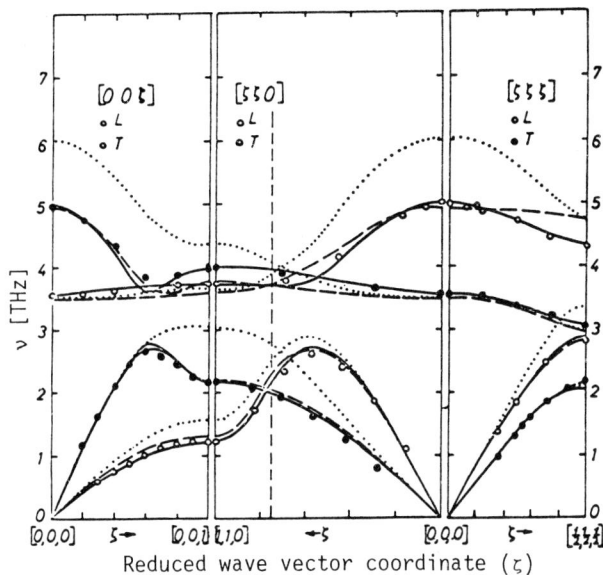

Fig. 4.13a. KBr: ω(q) [Ref. 4.36, Fig. 2], T = 90 K, Measurements [4.37], M: 7P-BSM (solid-lines), 7P-SM (dashed lines), 5P-FCM (dotted lines), Lit. [4.25,38,49,51-54]

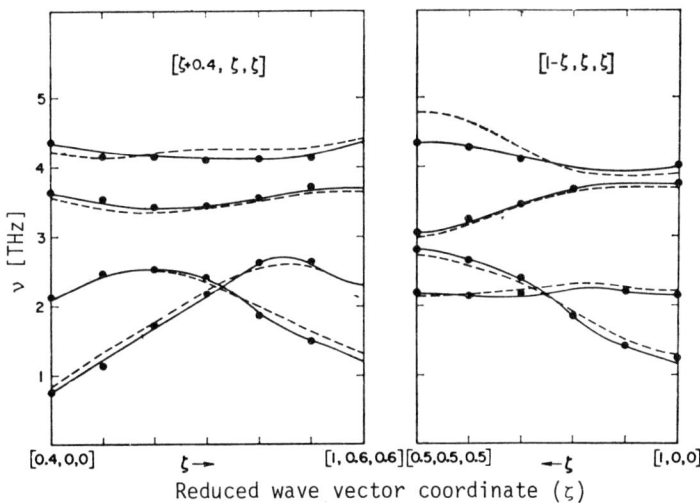

Fig. 4.13b. KBr: ω(q) [Ref. 4.51, Fig. 3], T = 90 K, M: 9P-SM (solid lines), 4P-SM (dashed lines), Lit. [4.25,38,39,49,51-54]

KBr

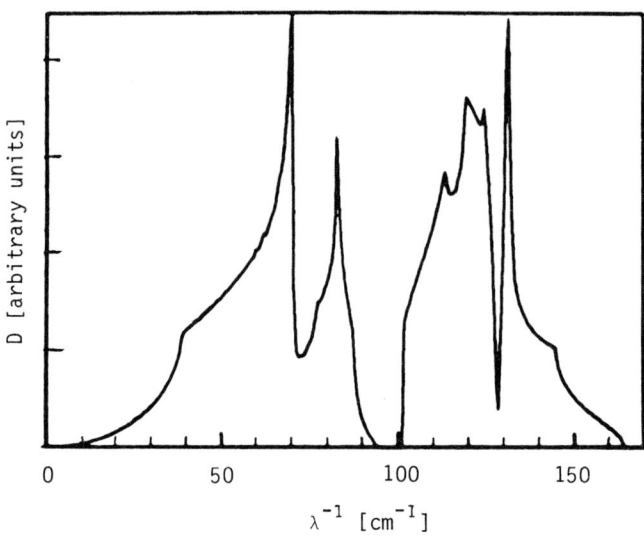

Fig. 4.13c. KBr: D(ω) [Ref. 4.25, Fig. 4], M: 11P-SM, Lit. [4.51]

KI

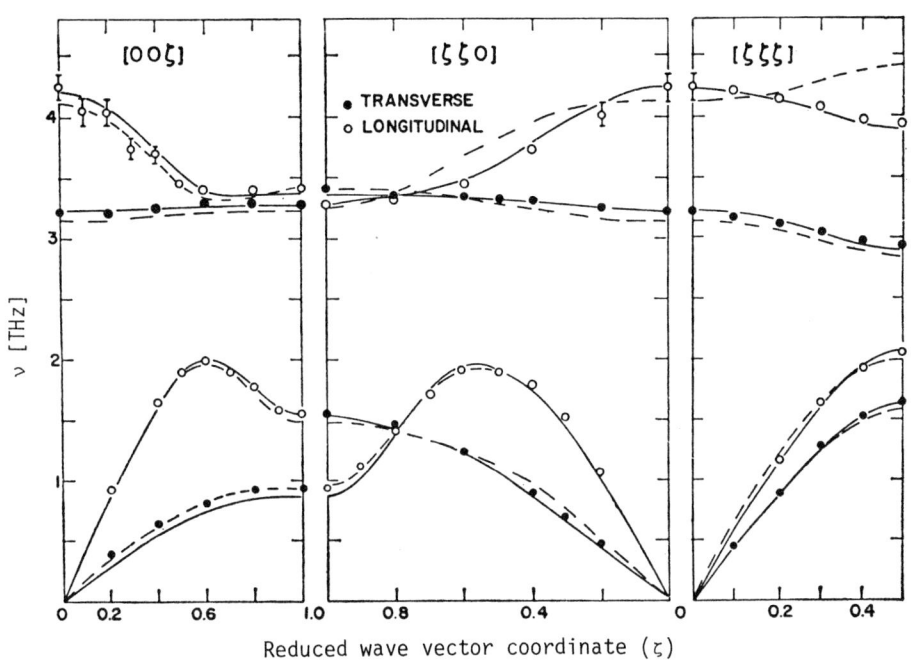

Fig. 4.14a. KI: $\omega(\underline{q})$ [Ref. 4.55, Fig. 1], T = 95 K, M: 11P-SM (solid lines), 3P-SM (dashed lines), Lit. [4.25,35,54]

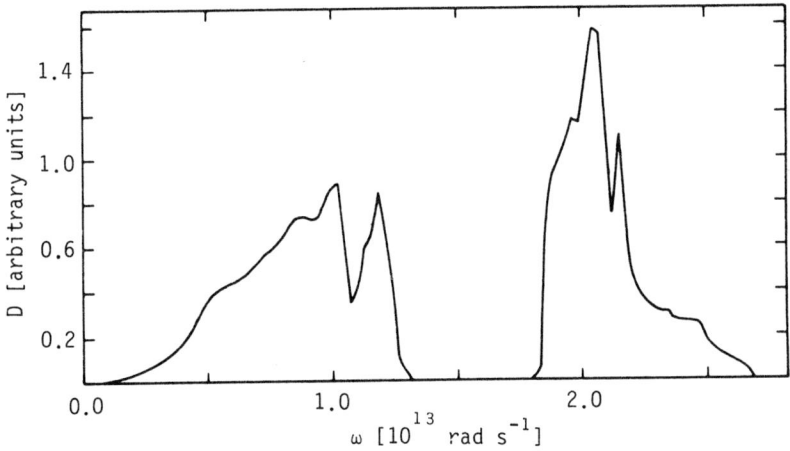

Fig. 4.14b. KI: $D(\omega)$ [Ref. 4.56, Fig. 9], T = 95 K, M: 11P-SM, Lit. [4.25]

RbF

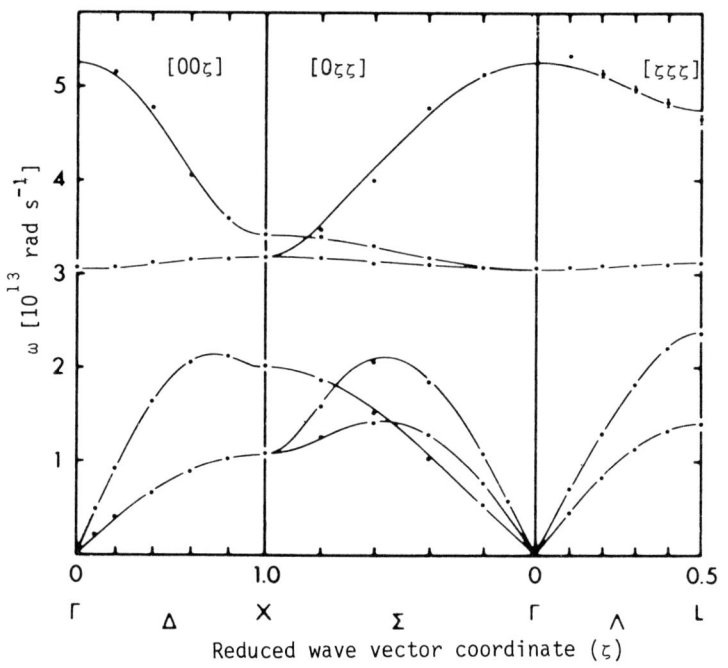

Fig. 4.15a. RbF: $\omega(\underline{q})$ [Ref. 4.26, Fig. 1d], T = 80 K, M: 11P-SM, Lit. [4.57,58,65]

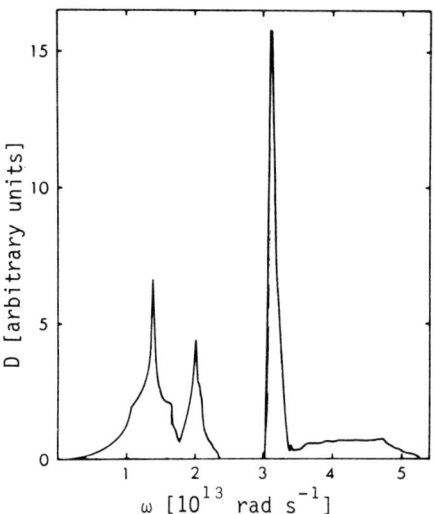

Fig. 4.15b. RbF: $D(\omega)$ [Ref. 4.26, Fig. 2d], T = 80 K, M: 11P-SM

RbCl

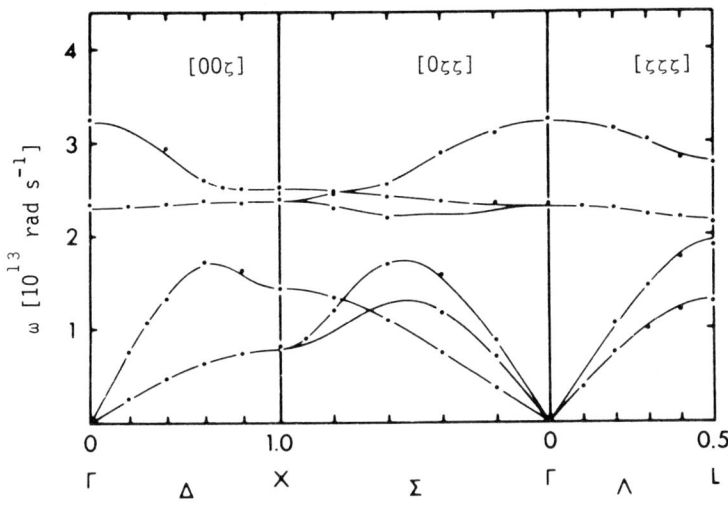

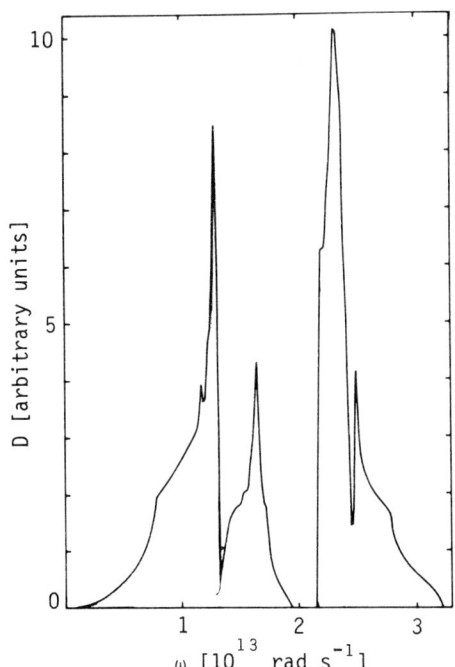

Fig. 4.16a. RbCl: $\omega(\underline{q})$ [Ref. 4.26, Fig. 1c], T = 80 K, M: 11P-SM, Lit. [4.57-59,65]

Fig. 4.16b. RbCl: $D(\omega)$ [Ref. 4.26, Fig. 2c], T = 80 K, M: 11P-SM

RbBr

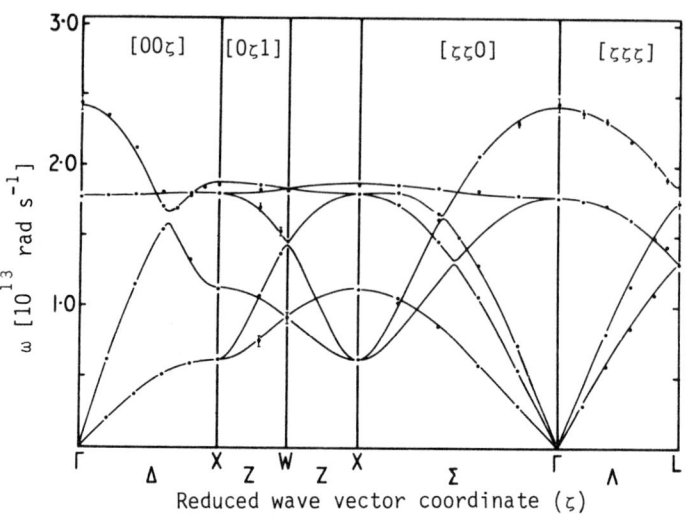

Fig. 4.17a. RbBr: ω(q) [Ref.4.60, Fig.1], T = 80 K, M: 11P-SM, Lit.[4.58,65]

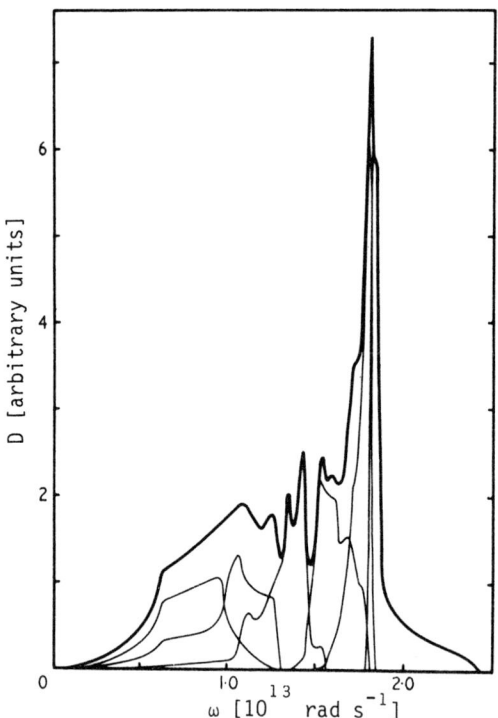

Fig. 4.17b. RbBr: D(ω) [Ref.4.60, Fig.2], T = 80 K, M: 11P-SM

RbI

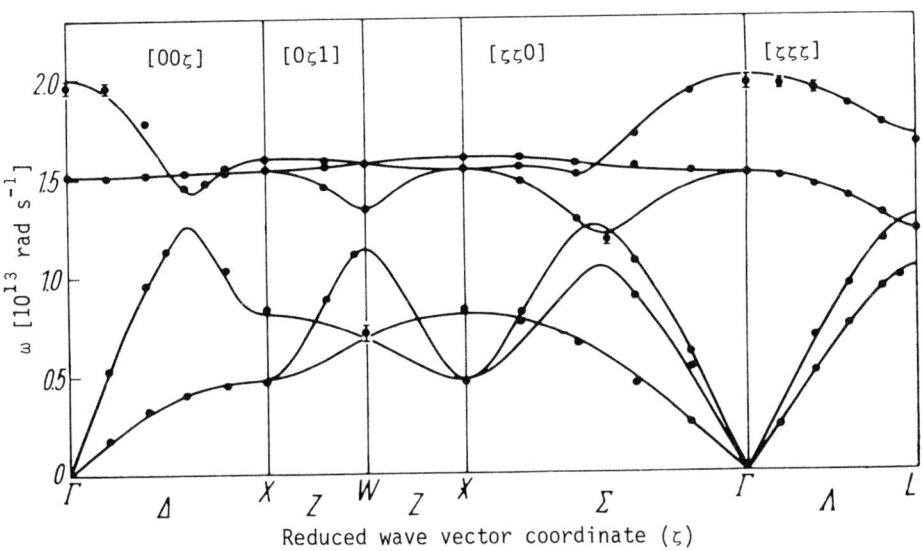

Fig. 4.18a. RbI: $\omega(\underline{q})$ [Ref.4.61, Fig.1], T = 80 K, M: 11P-SM, Lit. [4.58,65]

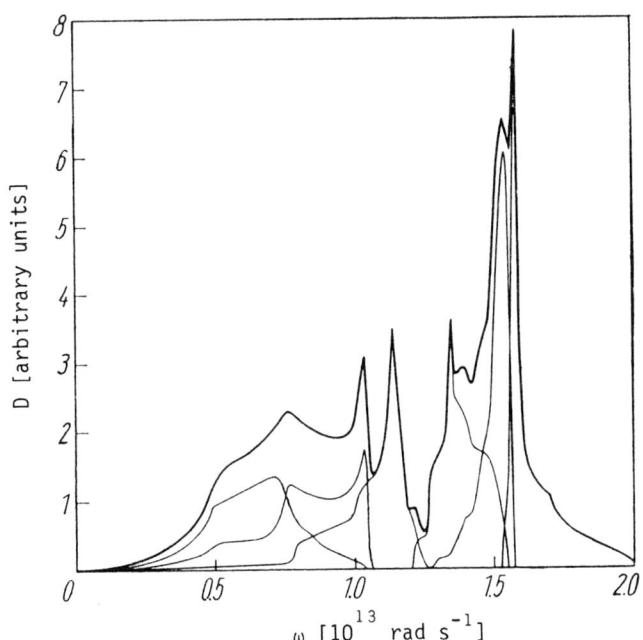

Fig. 4.18b. RbI: $D(\omega)$ [Ref.4.61, Fig.2], T = 80 K, M: 11P-SM

5. Metal Oxides (Rock Salt Structure)

These crystals show a behavior very similar to that of alkali halides. Shell models with about ten parameters give generally good agreement with experimental data. Of particular interest is CoO because of its phase transition from the antiferromagnetic to the paramagnetic state.

The following systems are treated:

Crystal	Figures showing	
	Dispersion curve $\omega(\underline{q})$	Density of states $D(\omega)$
MgO	5.1a	5.1b
CaO	5.2a	5.2b
SrO	5.3a	5.3b
BaO	5.4a	5.4b
MnO	5.5	
FeO	5.6a	5.6b
CoO (paramagnetic)	5.7a	
CoO (antiferromagnetic)	5.7b	
NiO	5.8a	5.8b

MgO

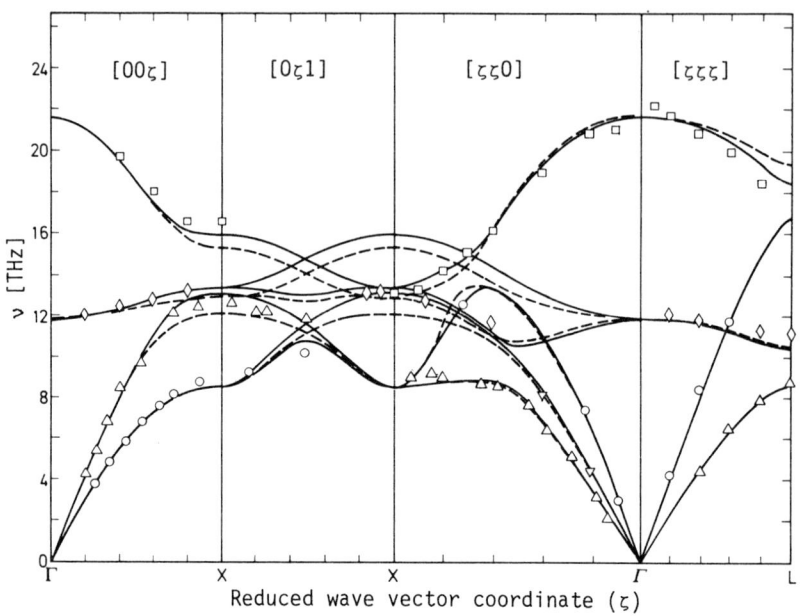

Fig. 5.1a. MgO: ω(q) [Ref. 5.1, Fig. 2], T = 293 K, M: 8P-BSM (full lines), 7P-SM (dashed lines), Lit. [5.2-7, 5.25]

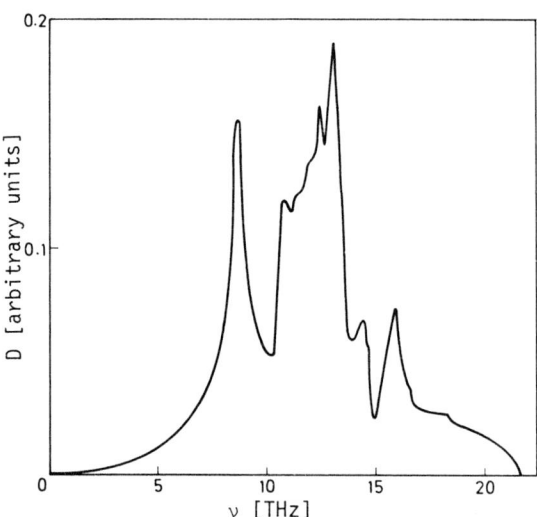

Fig. 5.1b. MgO: D(ω) [Ref. 5.1, Fig. 3], M: 8P-BSM, Lit. [5.26]

CaO

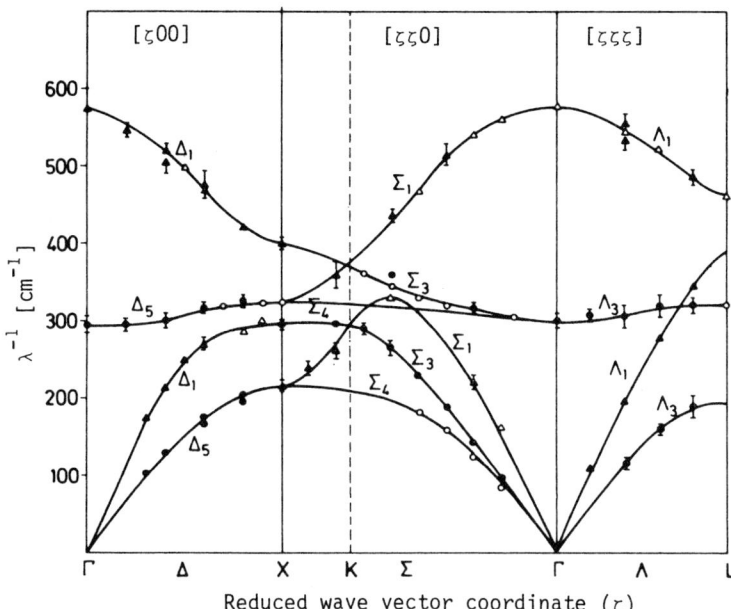

Fig. 5.2a. CaO: $\omega(\underline{q})$ [Ref. 5.7, Fig. 2], T = 293 K, full circles and triangles [5.8], open circles and triangles [5.9], M: 11P-SM, Lit. [5.10]

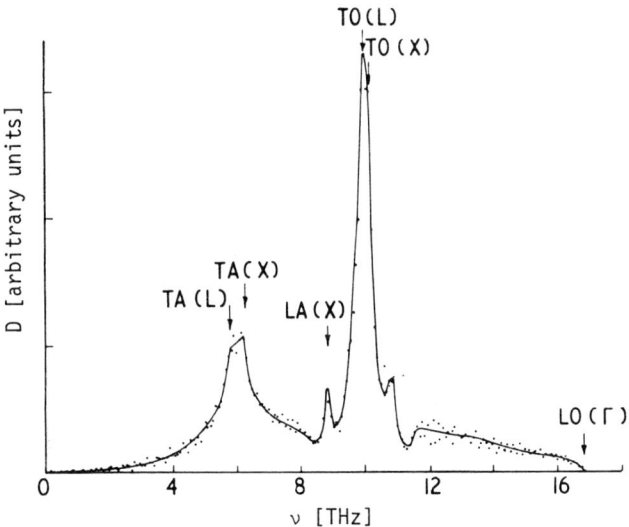

Fig. 5.2b. CaO: $D(\omega)$ [Ref. 5.8, Fig. 2], T = 293 K, M: 11P-SM, Lit. [5.9]

SrO

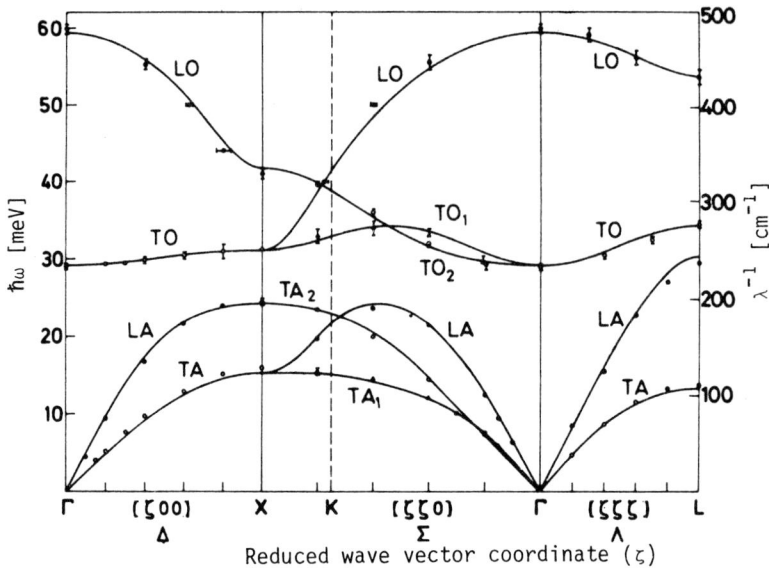

Fig. 5.3a. SrO: ω(q) [Ref. 5.11, Fig. 1], T = 300 K, M: 7P-SM, Lit.[5.10,12]

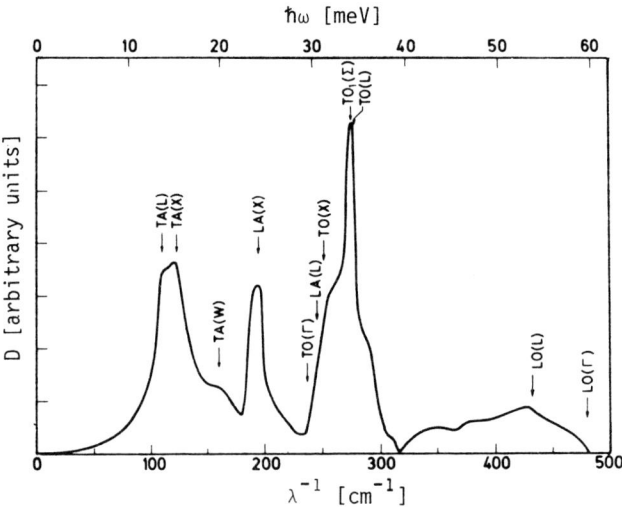

Fig. 5.3b. SrO: D(ω) [Ref. 5.11, Fig. 2], T = 300 K, M: 7P-SM

BaO

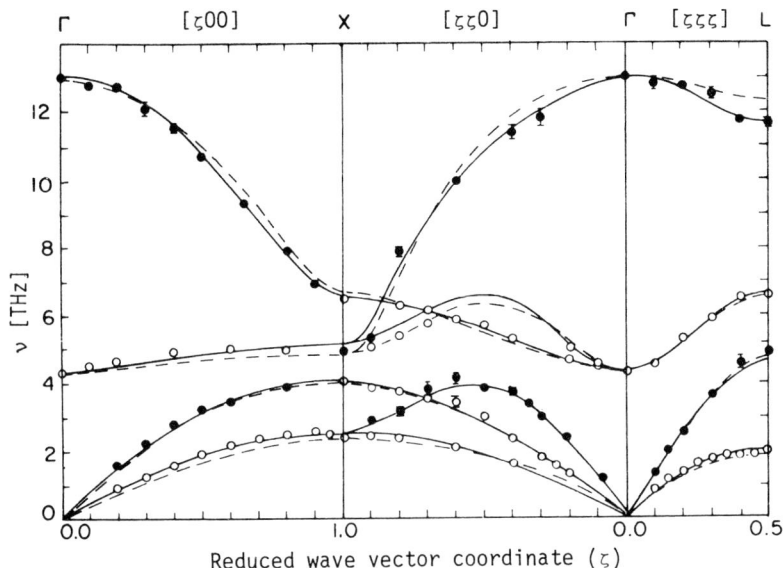

Fig. 5.4a. BaO: $\omega(\underline{q})$ [Ref. 5.13, Fig. 1], T = RT, M: 15P-SM (full lines), 8P-TSM (dashed lines)

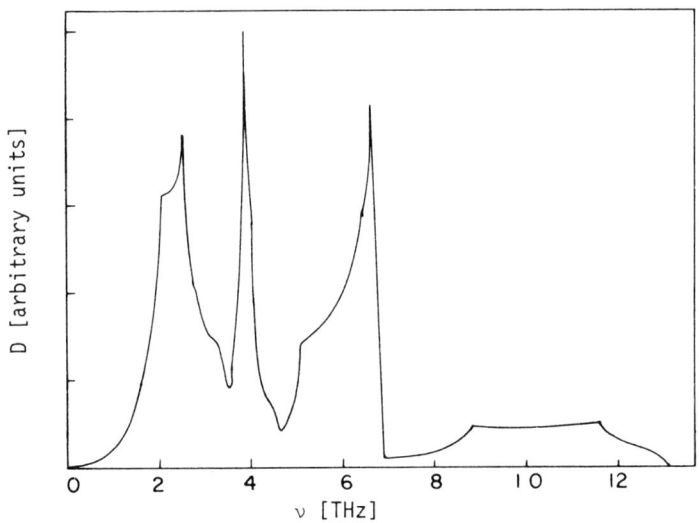

Fig. 5.4b. BaO: $D(\omega)$ [Ref. 5.13, Fig. 2], T = RT, M: 15P-SM

MnO

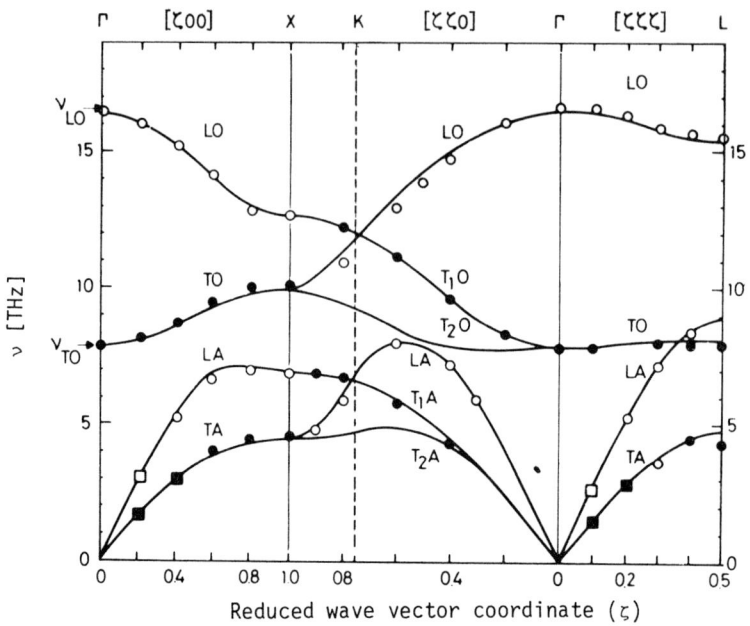

Fig. 5.5. MnO: ω(q) [Ref. 5.17, Fig. 1], T = RT, M: 9P-OSM, Lit. [5.15-18]

FeO

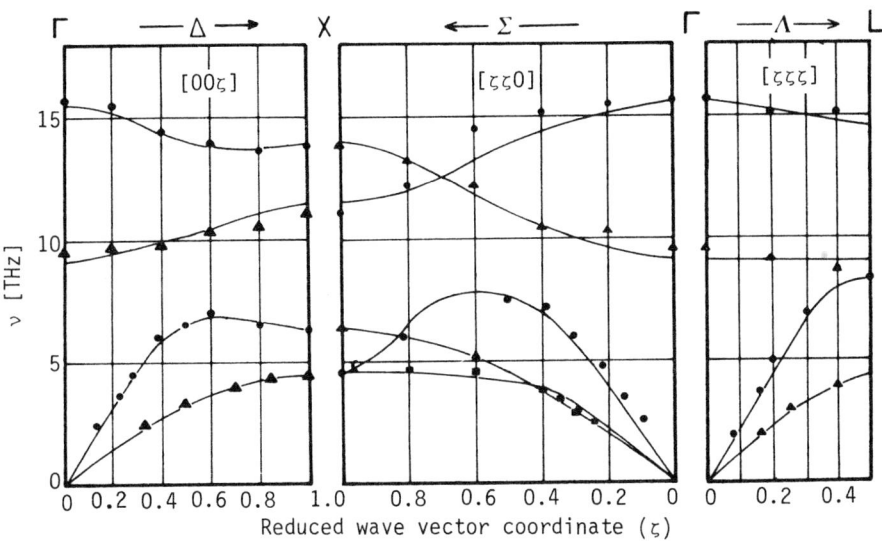

Fig. 5.6a. FeO: ω(q) [Ref. 5.19, Fig. 5], T = RT, M: 12P-SM, Lit. [5.14]

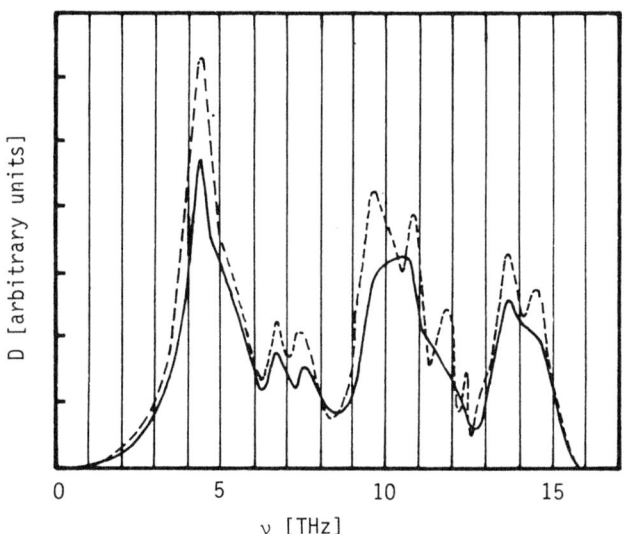

Fig. 5.6b. FeO: D(ω) [Ref. 5.19, Fig. 9], M: 12P-SM, two different channel widths

CoO (paramagnetic)   CoO (antiferromagnetic)

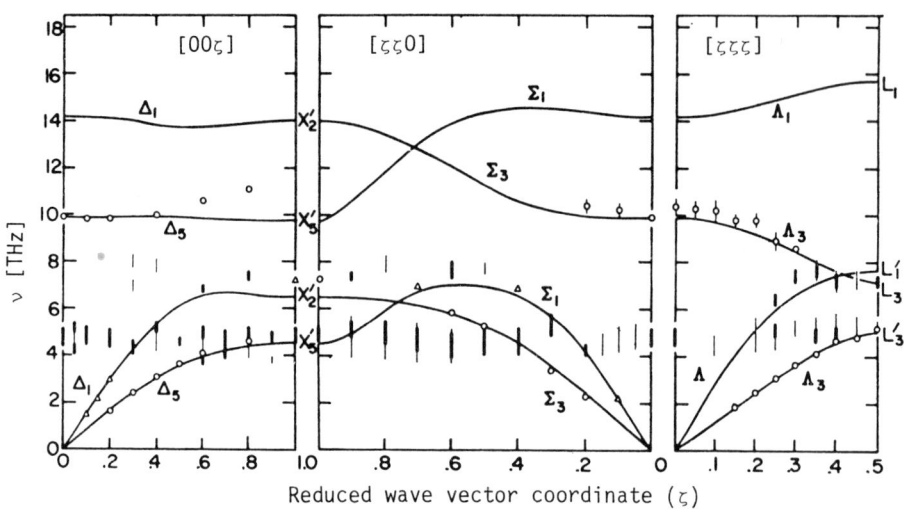

Fig. 5.7a. CoO (paramagnetic): $\omega(\underline{q})$ [Ref. 5.20, Fig. 5], T = 425 K for [00ξ] and [$\xi\xi$0], T = 330 K for [$\xi\xi\xi$], M: 11P-SM, Lit. [5.18,21]

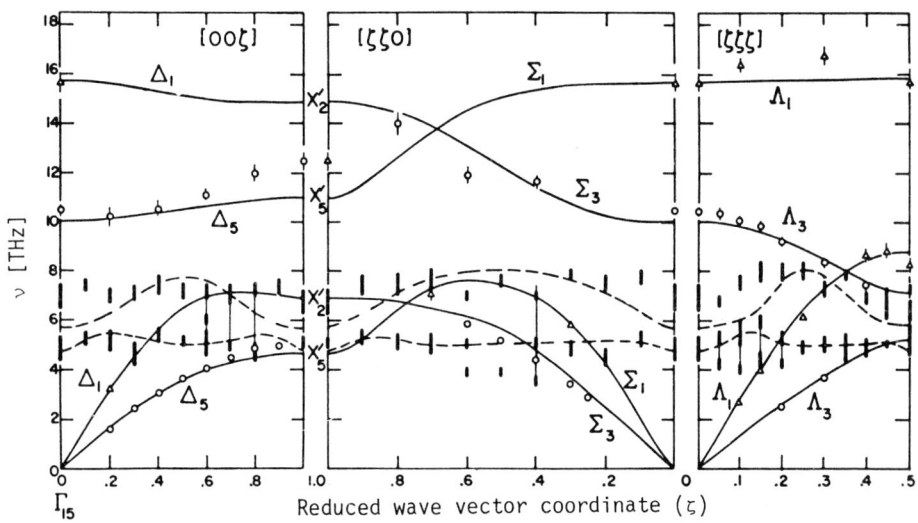

Fig. 5.7b. CoO (antiferromagnetic): $\omega(\underline{q})$ [Ref. 5.20, Fig. 4], T = 110 K M: 11P-SM, Lit. [5.18,21]

NiO

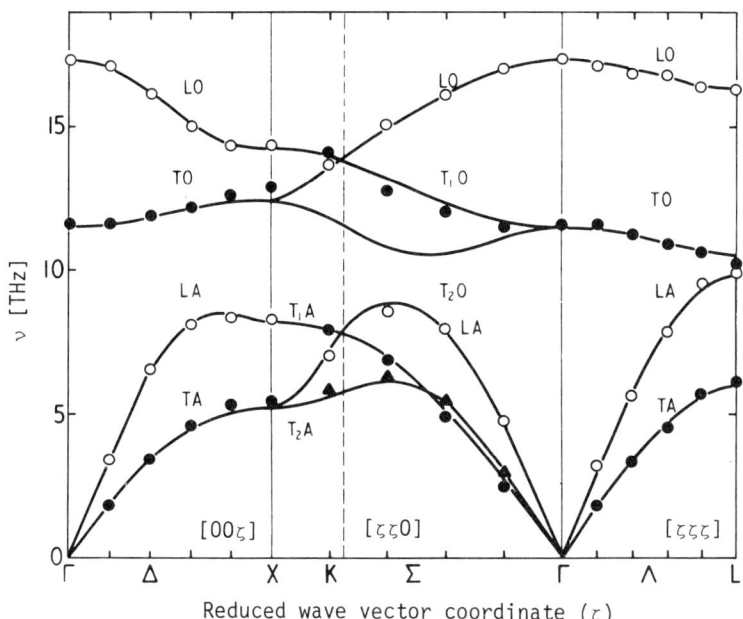

Fig. 5.8a. NiO: ω(q) [Ref. 5.22, Fig. 2], T = 293 K, M: 9P-SM, Lit. [5.14,18,23,24]

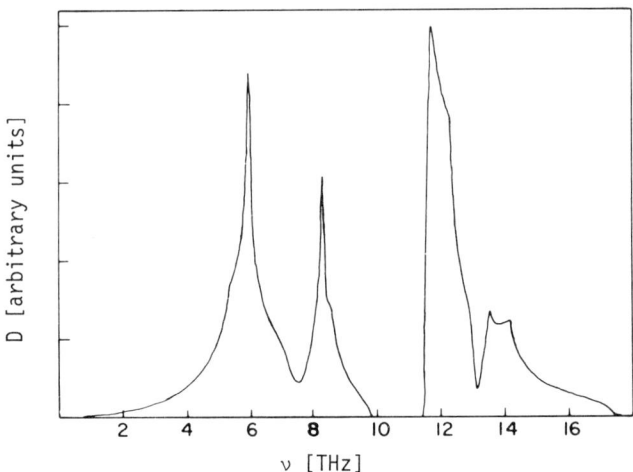

Fig. 5.8b. NiO: D(ω) [Ref. 5.23, Fig. 2], T = 297 K, M: 11P-SM

6. Transition Metal Compounds (Rock Salt Structure)

The interest in these groups of materials is related to the fact that many of them (the group V_b carbides and the group IV_b nitrides) are high-temperature superconductors. The general behavior of the phonon dispersion curves is still similar to those in other rock salt crystals and may well be described in terms of a shell model. Characteristic deviations appear in the optic modes at long wavelengths where the metallic screening destroys the Lyddane-Sachs-Teller splitting between the LO and TO modes, while at higher q values the Coulomb splitting between these modes can be clearly recognized. While this behavior is typical for all metallic crystals with optic modes, the phonon anomalies at q vectors with components $q_\alpha = 1/2$ (in particular LA (1/2,0,0) and TA,LA at the L point (1/2,1/2,1/2) seem to correlate with high T_c's in a way typical for this family of compounds. A successful model description was given by WEBER in his DSM [6.6,13], while different microscopic explanations were given in [6.19,25].

Crystal	Figures showing		Crystal	Figures showing	
	Dispersion curve $\omega(\underline{q})$	Density of states $D(\omega)$		Dispersion curve $\omega(\underline{q})$	Density of states $D(\omega)$
YS	6.1a	6.1b	NbC_{1-x}	6.8a,b	6.8c
TiC	6.2a	6.2b	NbN (γ-)	6.9a	
TiN	6.3a	6.3b	NbN (δ-)		6.9b
ZrC	6.4a	6.4b	TaC	6.10	
ZrN		6.5	UC	6.11	
HfC	6.6		UN	6.12	
VC	6.7a	6.7b			

Note added in proof:

The phonon dispersion curves of ZrN have recently been measured [A.N. Christensen, O.W. Dietrich, W. Kress, W.D. Teuchert: Phys. Rev. B *19*, 5699 (1979)]

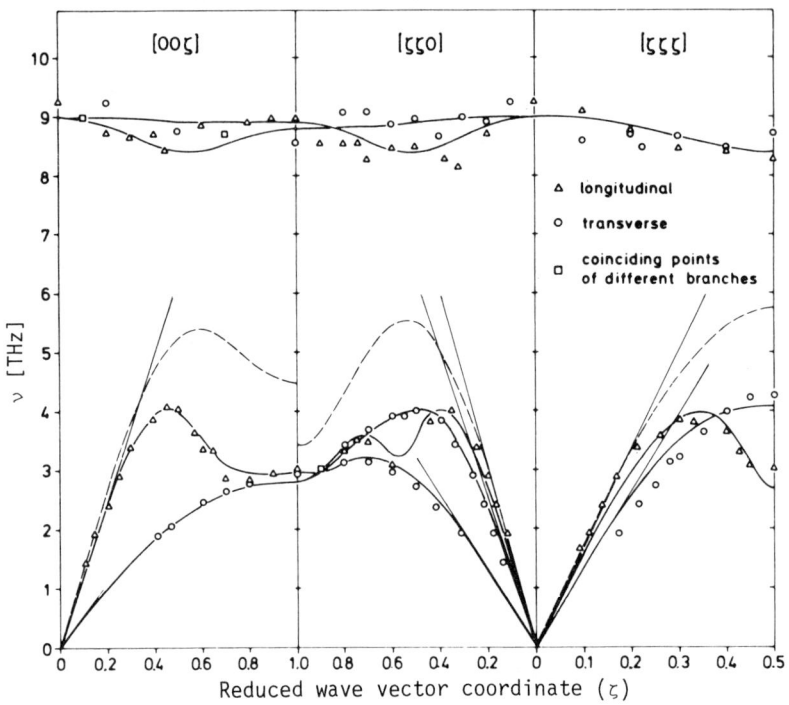

Fig. 6.1a. YS: ω(q̄) [Ref. 6.1, Fig. 1], T = RT, M: 11P-DSM (full lines), 5P-SM (dashed lines), Lit. [6.2]

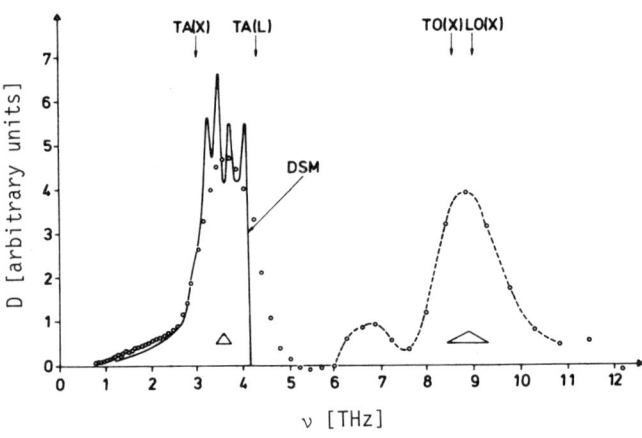

Fig. 6.1b. YS: D(ω) [Ref. 6.2, Fig. 2], T = RT, M: 11P-DSM (full lines), guide lines to the eye (dashed lines)

TiC

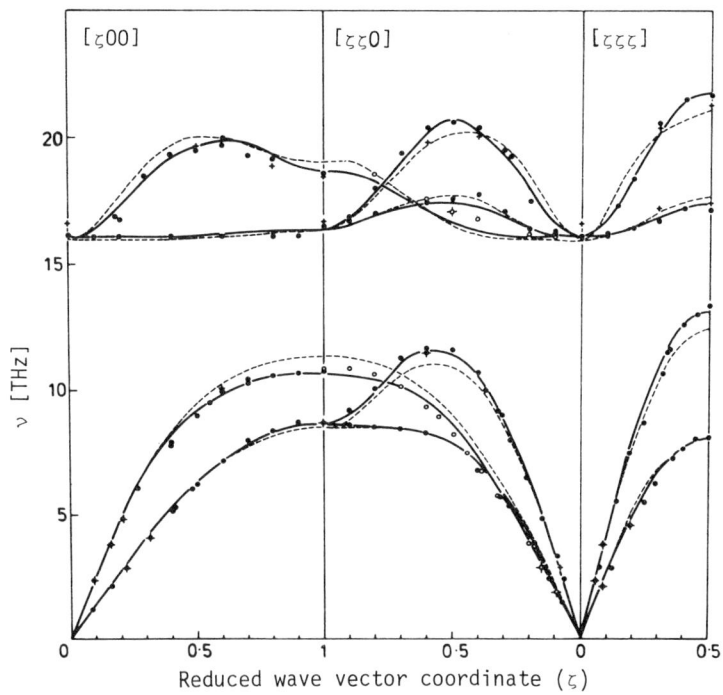

Fig. 6.2a. TiC: $\omega(\underline{q})$ [Ref. 6.3, Fig. 1], T = RT, circles $TiC_{0.95}$, crosses $TiC_{0.89}$, M: 12P-SM (full lines), 8P-SM (dashed lines)

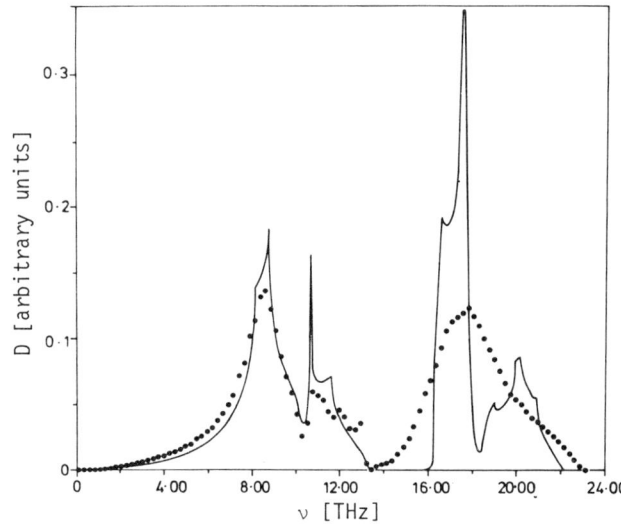

Fig. 6.2b.
$TiC_{0.95}$: $D(\omega)$
[Ref. 6.3, Fig. 2]
T = RT, M: 12P-SM

TiN

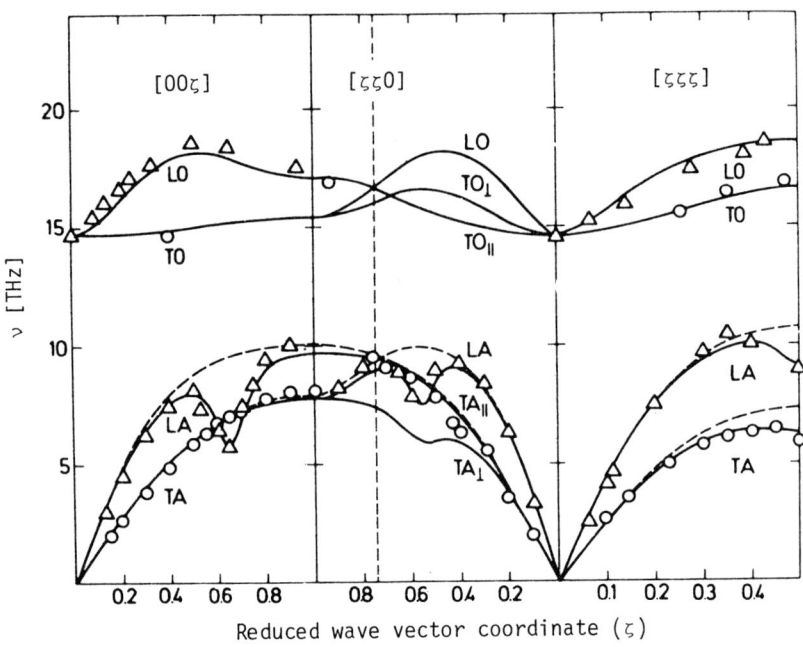

Fig. 6.3a. TiN$_{0.98}$: $\omega(\underline{q})$ [Ref. 6.4, Fig. 1], T = RT, M: 13P-DSM (full lines), 7P-SM (dashed lines)

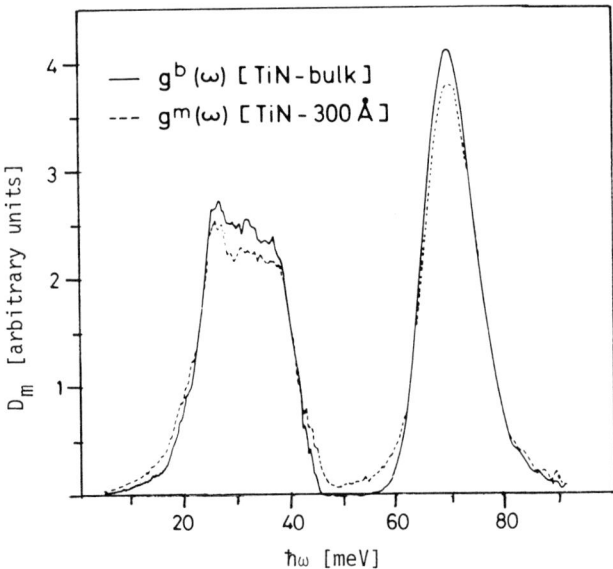

Fig. 6.3b. TiN: $D(\omega)$ [Ref. 6.5, Fig. 2a], T = 296 K, M: -, bulk (full lines), microcrystalline (dashed lines), Lit. [6.4]

ZrC

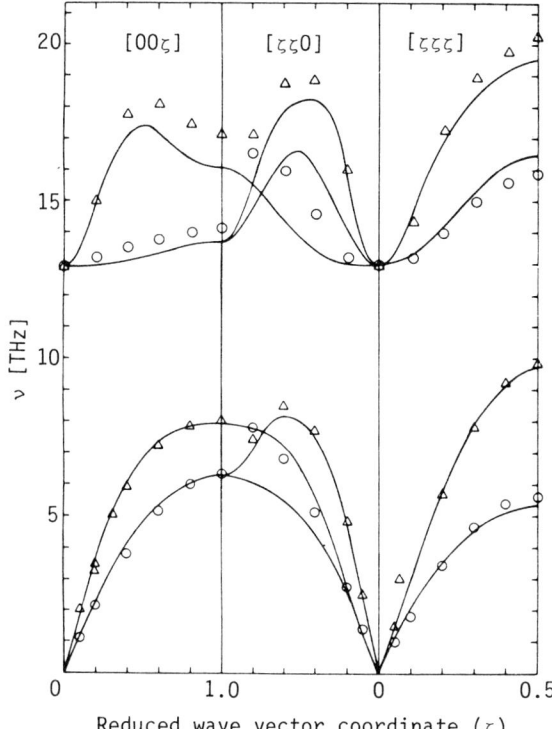

Fig. 6.4a. ZrC: $\omega(\underline{q})$ [Ref. 6.6, Fig. 6], T = 298 K, M: 9P-SM, measurements [6.7,8], Lit. [6.9]

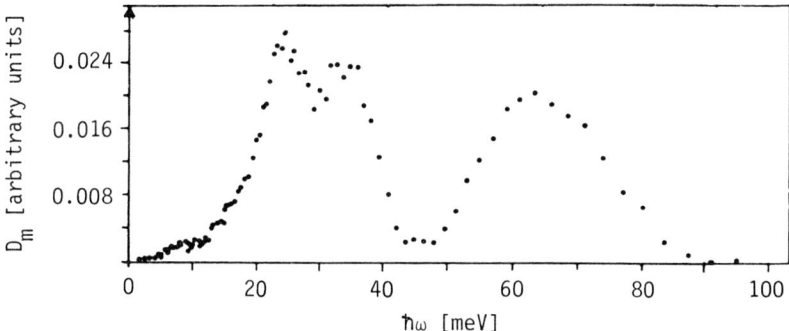

Fig. 6.4b. ZrC: $D(\omega)$ [Ref. 6.10, Fig. 13], T = RT

ZrN

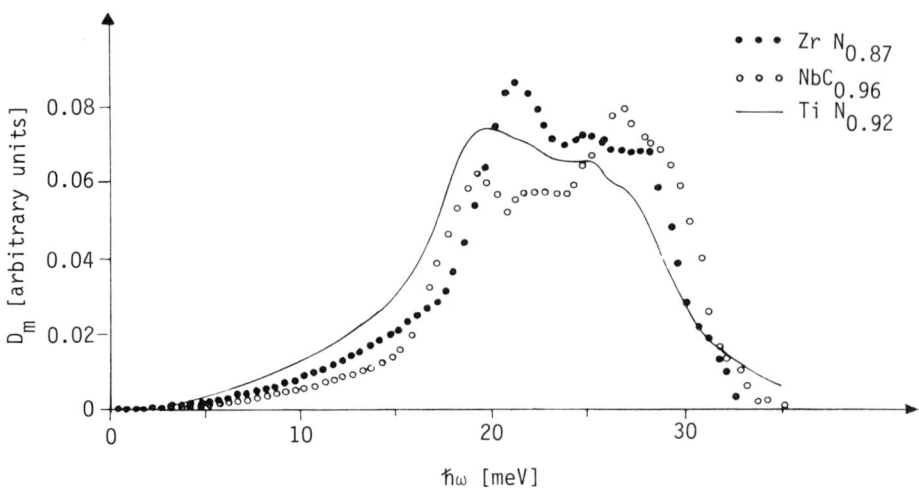

Fig. 6.5. ZrN: $D(\omega)$ [Ref. 6.11, Fig. 1], T = RT

HfC

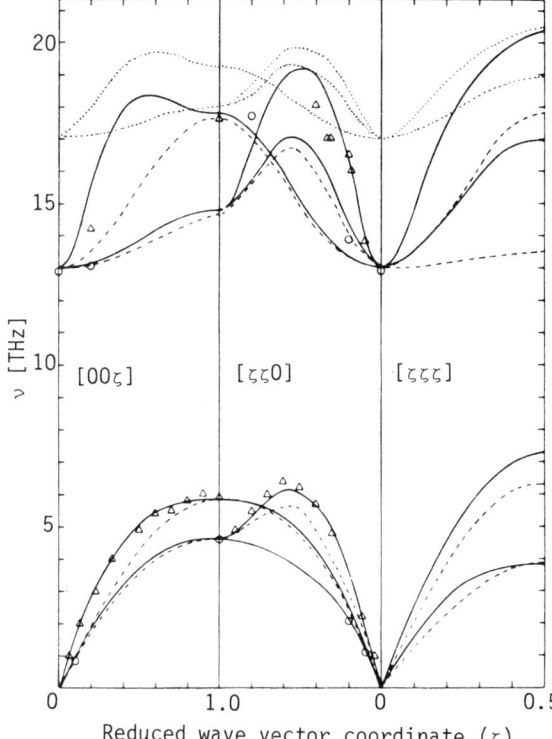

Fig. 6.6. HfC: ω(q)
[Ref. 6.6, Fig. 5],
T = 298 K, M: 9P-SM
measurements [6.12]
Lit. [6.13-15]

VC

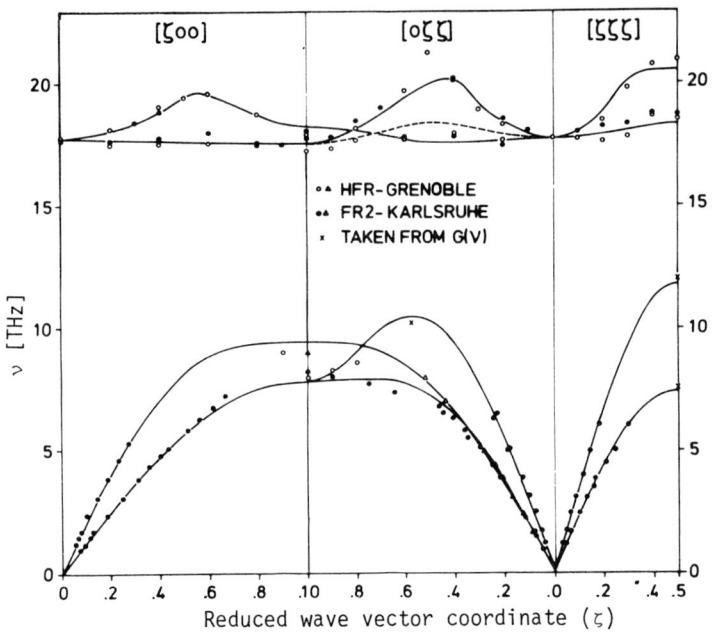

Fig. 6.7a. $VC_{0.87}$: $\omega(\underline{q})$ [Ref. 6.16, Fig. 1], T = RT, M: 9P-SM + 2P free electron screening

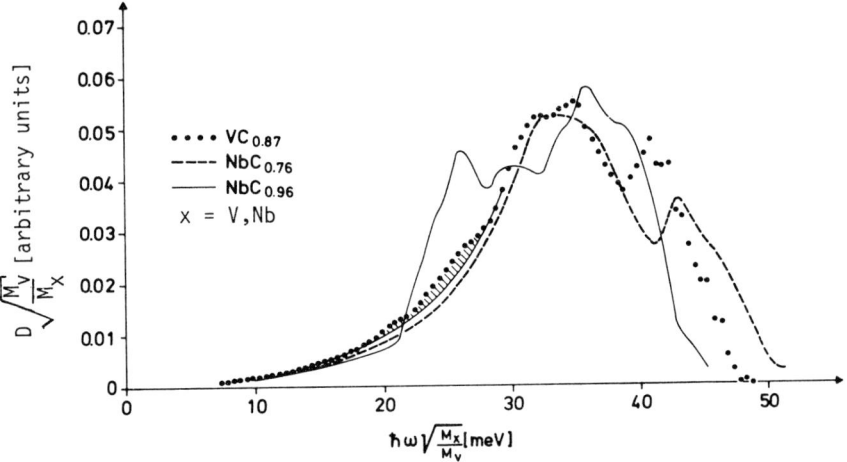

Fig. 6.7b. $VC_{0.87}$: $D(\omega)$ [Ref. 6.26, Fig. 1], T = RT

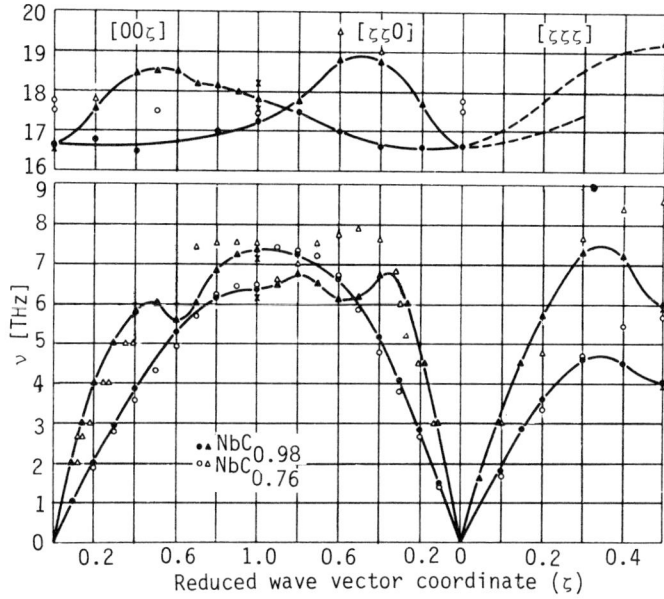

Fig. 6.8a. NbC_{1-x} $\omega(\underline{q})$ [Ref. 6.17, Fig. 3], T = 298 K, M: -, Lit. [6.6,18-21,25]

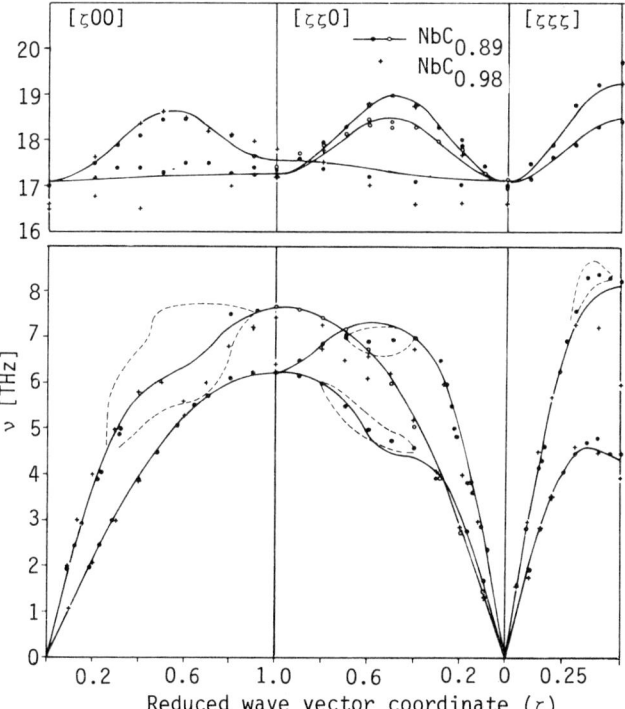

Fig. 6.8b. $NbC_{0.89}$: $\omega(\underline{q})$ [Ref. 6.18, Fig. 5], T = 298 K, data for $NbC_{0.98}$ [6.17], M: 14P-DSM (full lines), dashed lines: width of neutron groups, Lit. [6.6,17,19-21,25]

NbC

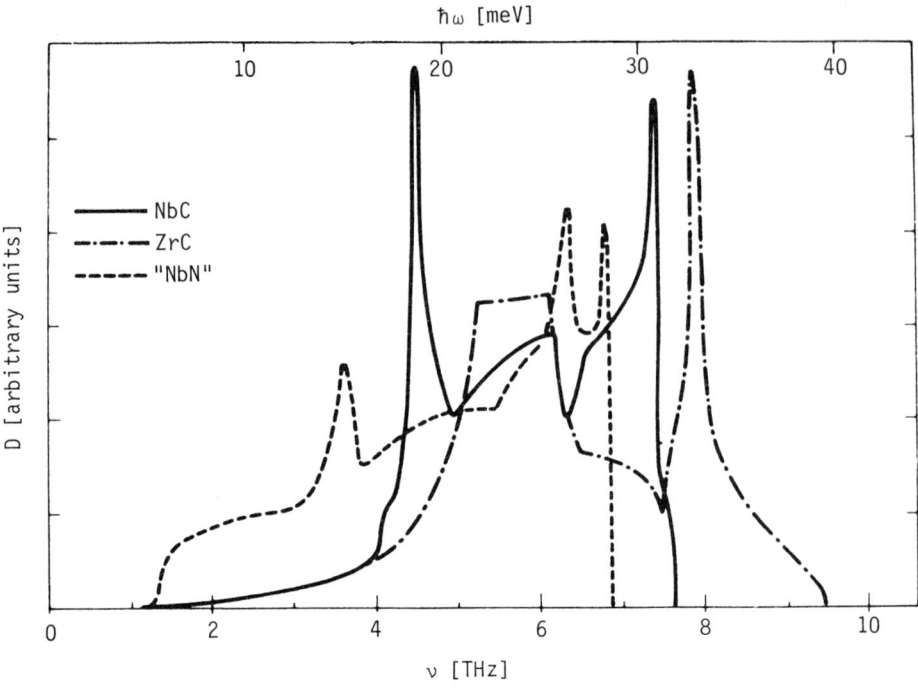

Fig. 6.8c. NbC: $D(\omega)$ [Ref. 6.6, Fig. 10], M: 14P-DSM, acoustic phonons only

γ-NbN　　　　　　　　　　　　　　　　　　　　　　δ-NbN

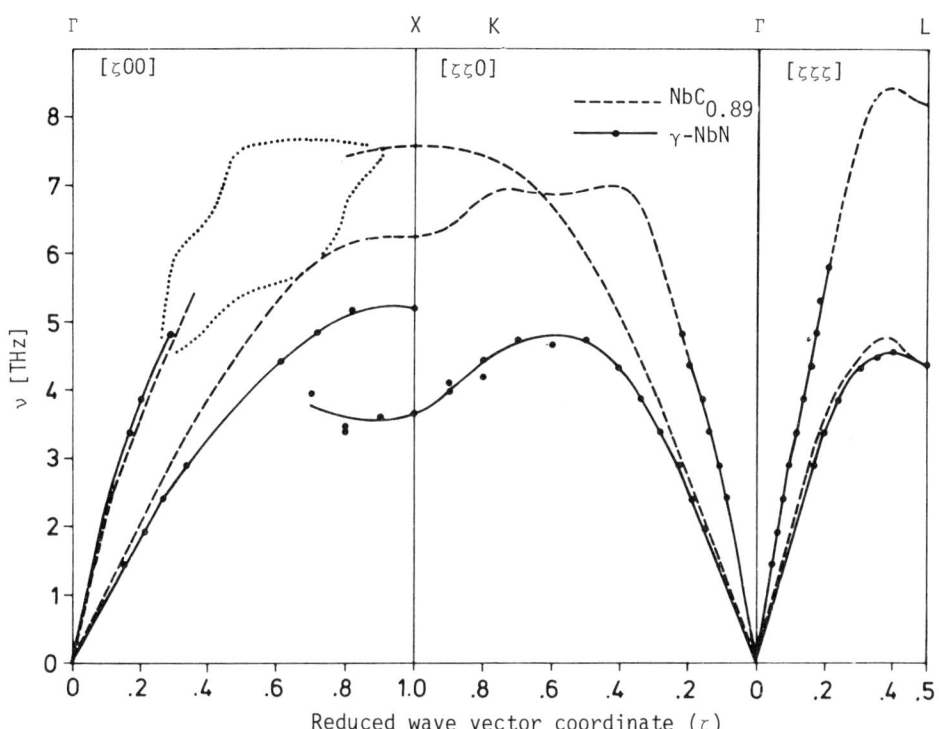

Fig. 6.9a. γ-NbN: $\omega(\underline{q})$ [Ref. 6.18, Fig. 1], M: -, Lit. [6.6]

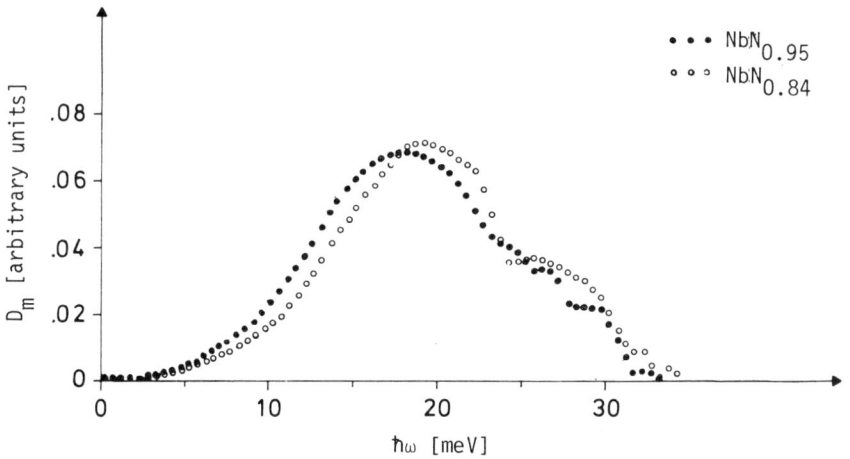

Fig. 6.9b. δ-NbN: $D(\omega)$ [Ref. 6.22, Fig. 1], M: -, Lit. [6.6]

TaC

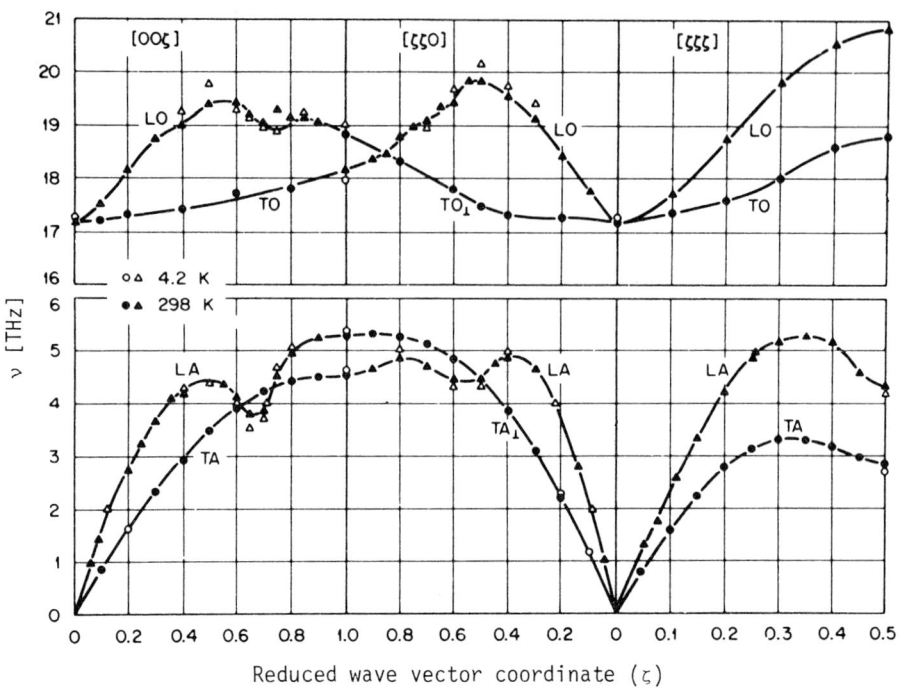

Fig. 6.10. TaC: $\omega(\underline{q})$ [Ref. 6.17, Fig. 1], T = 298 K, 4.2 K, M: -, Lit. [6.6,12-15]

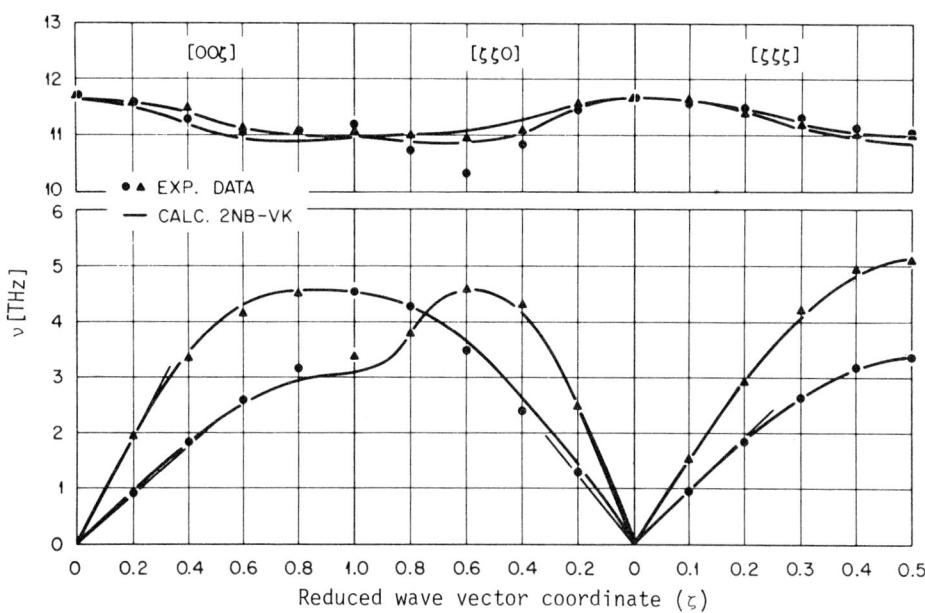

Fig. 6.11. UC: $\omega(\underline{q})$ [Ref. 6.7, Fig. 6], T = 298 K, M: 2NN-FCM, Lit. [6.6,14,23]

UN

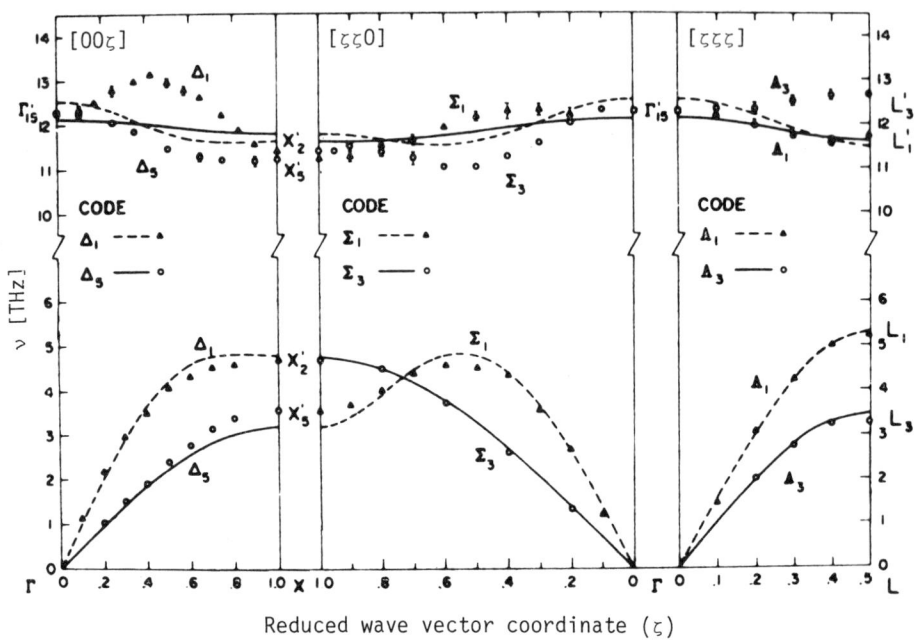

Fig. 6.12. UN: $\omega(q)$ [Ref. 6.24, Fig. 1], T = 4.2 K, M: 7P-RIM

7. Other Cubic Crystals (Rock Salt Structure)

This is physically the most interesting family of crystals among the "simple" cubic crystals. The first group (SnTe, etc.) exhibits interesting dielectric features due to the high polarizability of cations and anions. The silver halides show a characteristic deviation from other rock salt halides in the TO branch near all q points with components $q_\alpha \sim 1/2$. The origin of these anomalies is the quadrupolar deformability of the silver ion due to virtual electronic d → s excitations. The alkali cyanides behave like molecular crystals with (strongly temperature-dependent) rotational excitations and phase transitions.

The following systems are treated:

Crystal	Figures showing	
	Dispersion curve $\omega(\underline{q})$	Density of states $D(\omega)$
SnTe	7.1a	7.1b
PbS	7.2a	7.2b
PbSe	7.3a	7.3b
PbTe	7.4a	7.4b
NdSb	7.5a	7.5b
AgCl	7.6a	7.6b
AgBr	7.7a	7.7b
NaCN	7.8	
KCN	7.9a	7.9b
ND_4I	7.10a	7.10b
CsF	7.11a	7.11b

Note added in proof: The dispersion curves $\omega(\underline{q})$ of the SmS and $Sm_{0.75}Y_{0.25}S$ are found in [7.37] and [7.38] respectively.

SnTe

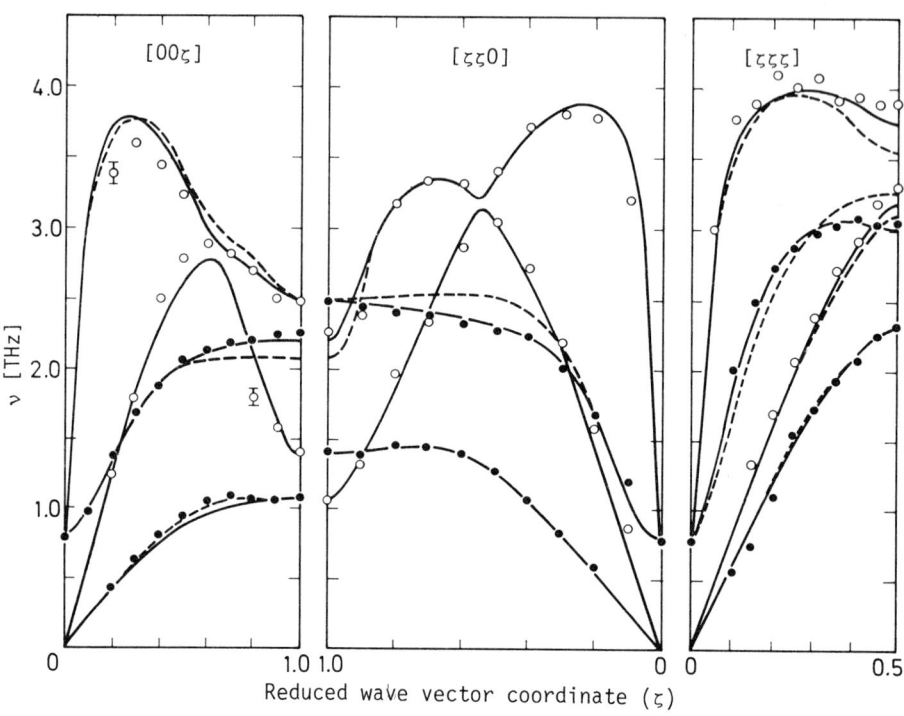

Fig. 7.1a. SnTe: ω(q) [Ref. 7.1, Fig. 1], T = 100 K, M: 15P-SM, full lines Z ≠ 0, dashed lines Z = 0, Lit.[7.2]

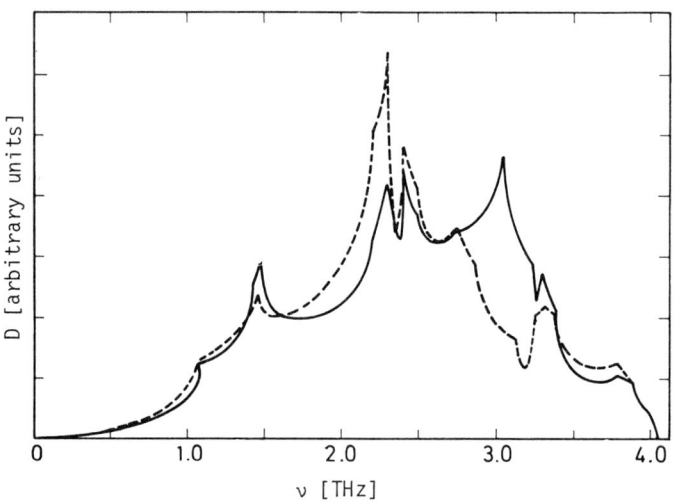

Fig. 7.1b. SnTe: D(ω) [Ref.7.1, Fig.2], full lines; dashed lines: diagonal element of the Green function, Lit.[7.2]

PbS

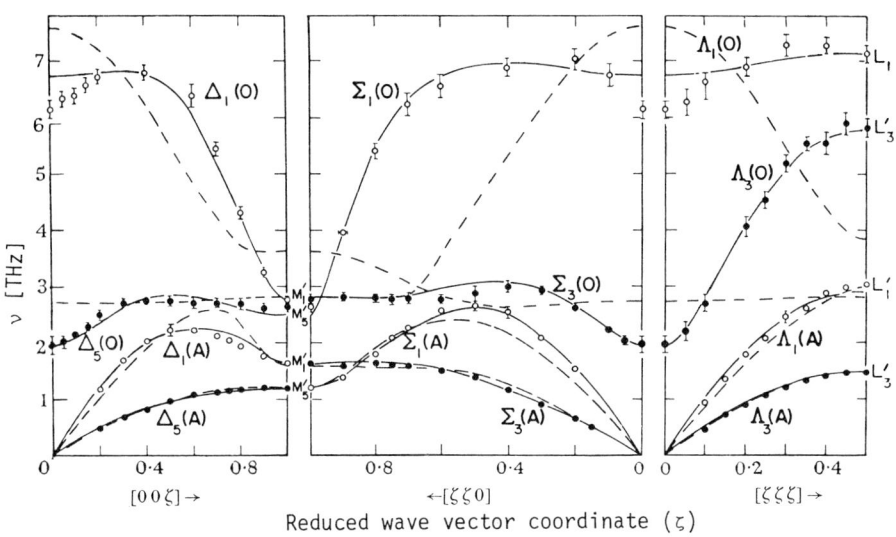

Fig. 7.2a. PbS: $\omega(\underline{q})$ [Ref. 7.3, Fig. 1], T = 296 K, M: 14P-SM (solid lines), 7P-FCM (dashed lines)

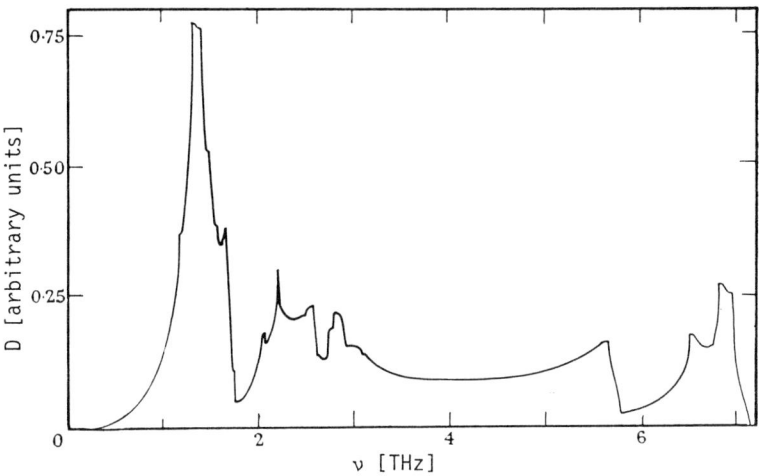

Fig. 7.2b. PbS: $D(\omega)$ [Ref. 7.3, Fig. 2], M: 14P-SM

PbSe

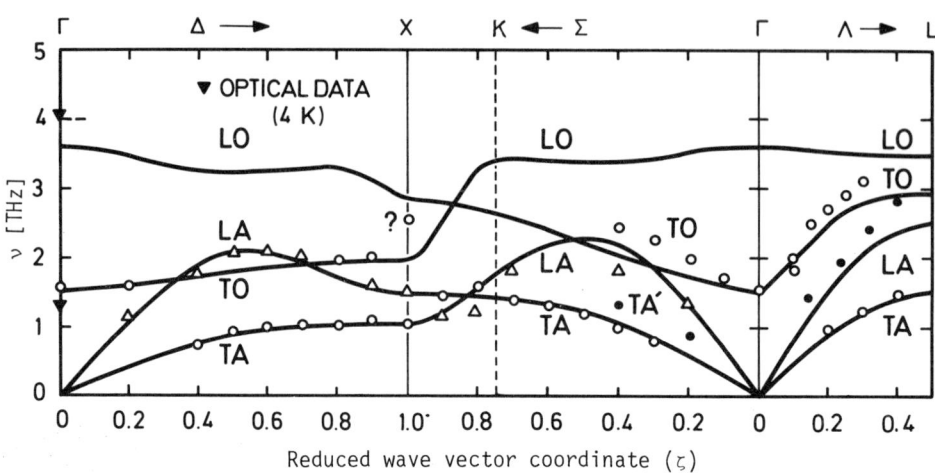

Fig. 7.3a. PbSe: ω(q) [Ref. 7.4, Fig. 1], T = 296 K, M: 7P-SM, Lit. [7.3,35]

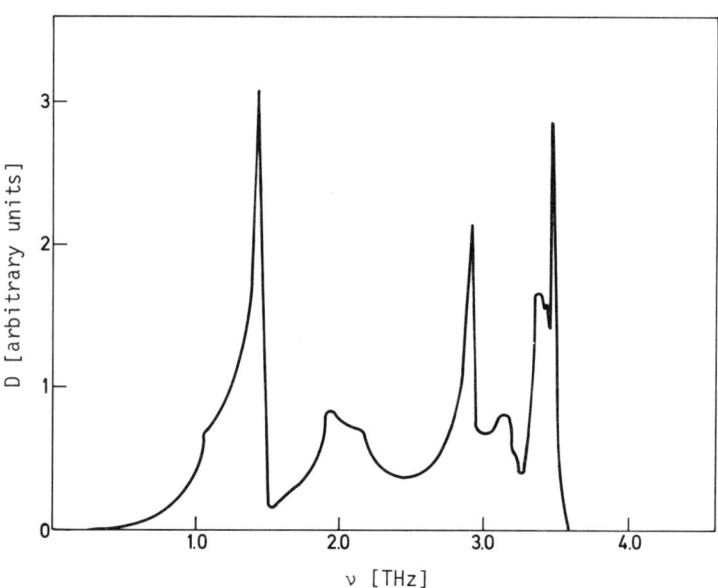

Fig. 7.3b. PbSe: D(ω) [Ref. 7.36], T = 296 K, M: 13P-SM, Lit. [7.3,35]

PbTe

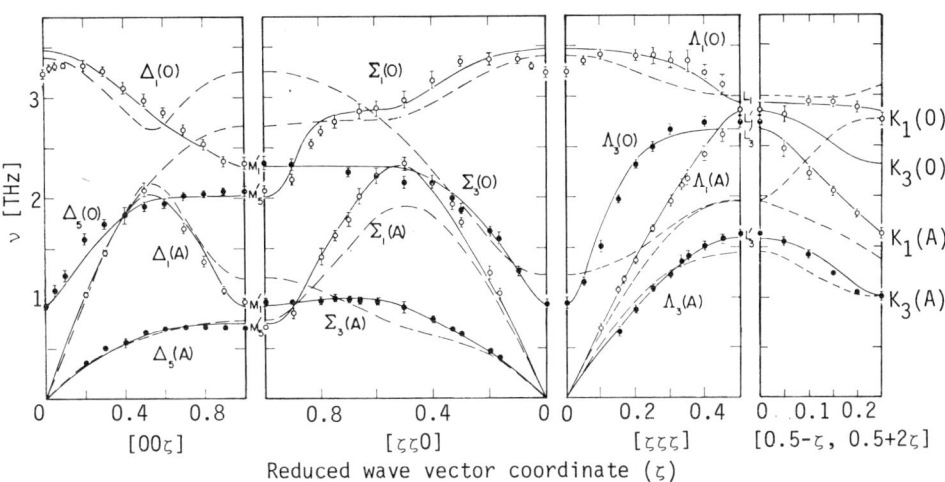

Fig. 7.4a. PbTe: ω(q) [Ref. 7.5, Fig. 1], T = 296 K, M: 13P-SM (full lines), 7P-FCM (dashed lines), Lit. [7.6]

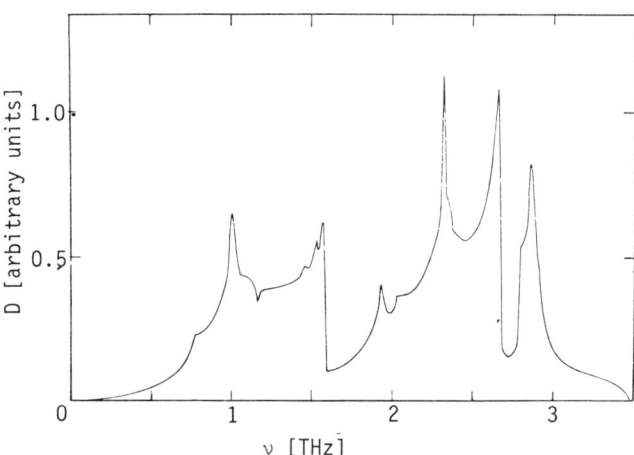

Fig. 7.4b. PbTe: D(ω) [Ref. 7.5, Fig. 2], M: 13P-SM (full lines)

NdSb

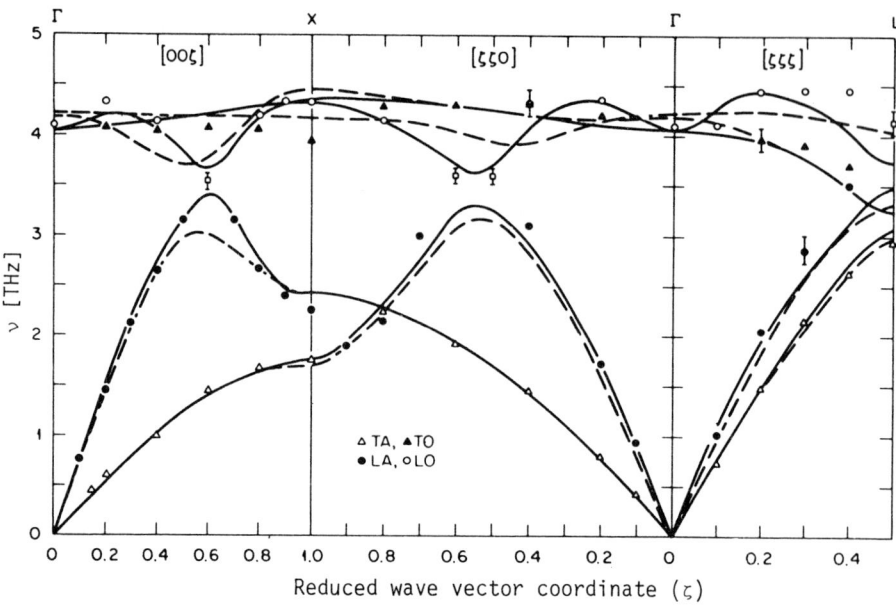

Fig. 7.5a. NdSb: $\omega(\underline{q})$ [Ref. 7.7, Fig. 1], T = RT, M: 9P- screened RIM (full lines), 2NN-SM (dashed lines)

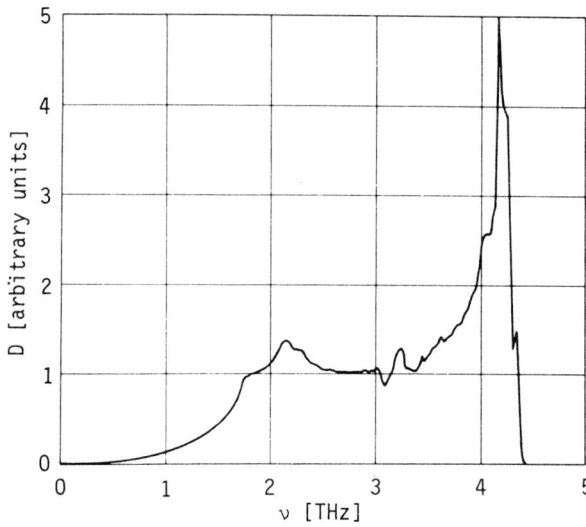

Fig. 7.5b. NdSb: $D(\omega)$ [Ref. 7.7, Fig. 2], M: 9P- screened RIM

AgCl

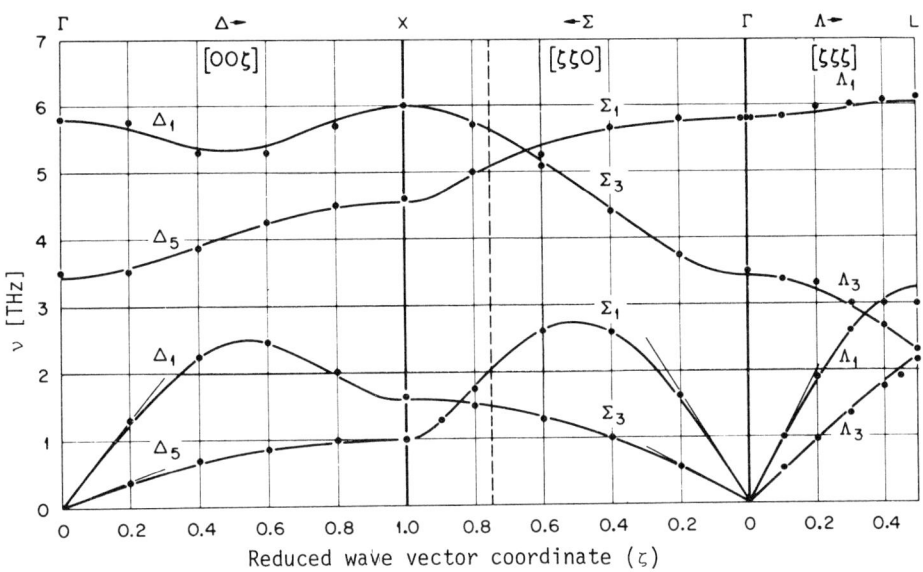

Fig. 7.6a. AgCl: $\omega(\underline{q})$ [Ref. 7.8, Fig. 1], T = 78 K, M: 13P-SM, Lit. [7.9-11]

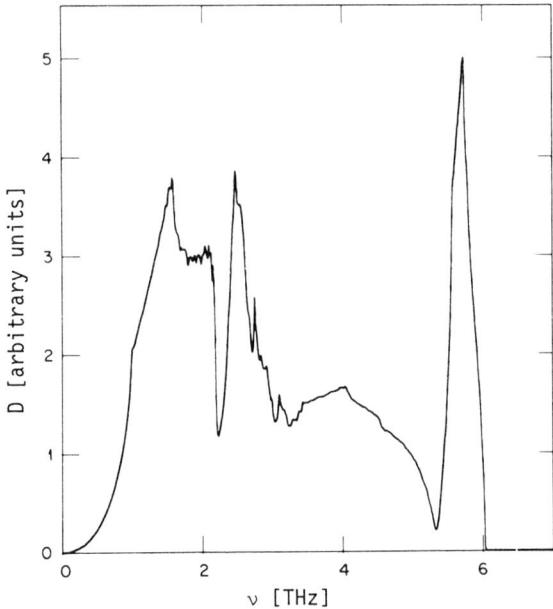

Fig. 7.6b. AgCl: $D(\omega)$ [Ref. 7.8, Fig. 3], T = 78 K, M: 13P-SM

AgBr

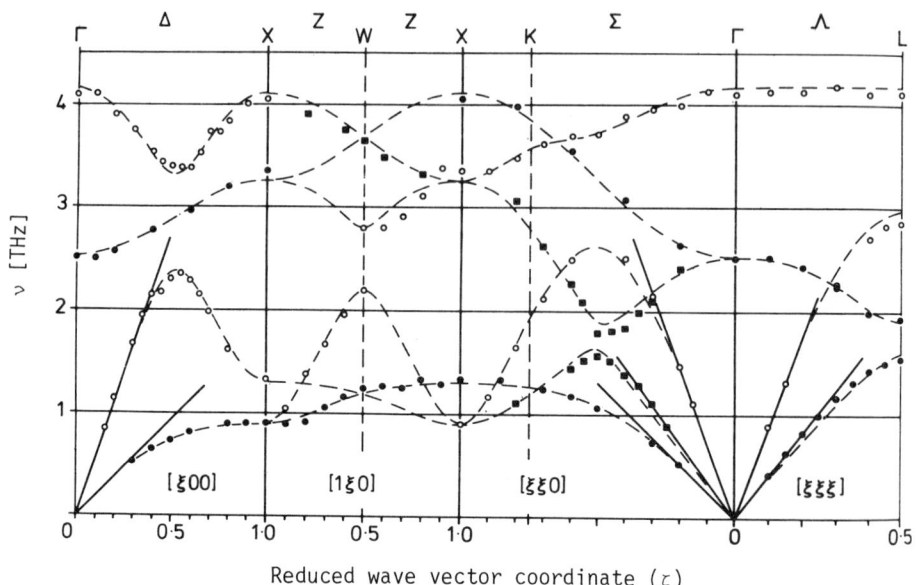

Fig. 7.7a. AgBr: ω(q) [Ref. 7.14, Fig. 1], T = 85 K, M: 15P-QSM, Lit. [7.10,12-20]

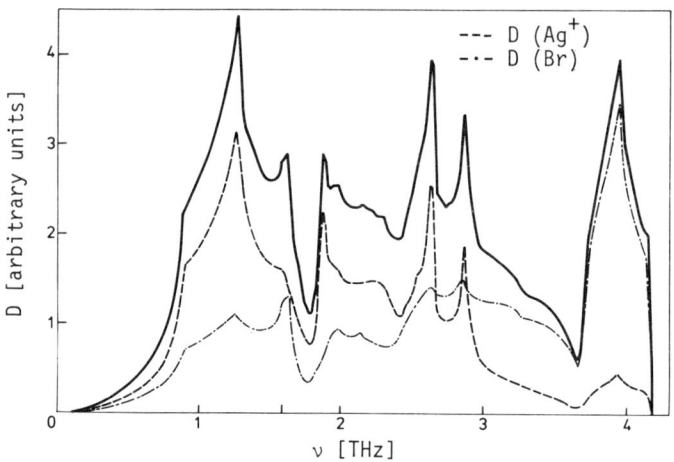

Fig. 7.7b. AgBr: D(ω) [Ref. 7.14, Fig. 4], T = 85 K, M: 15P-QSM, total density (full lines), partial densities of individual ions (dashed lines), Lit. [7.10,16-18]

NaCN

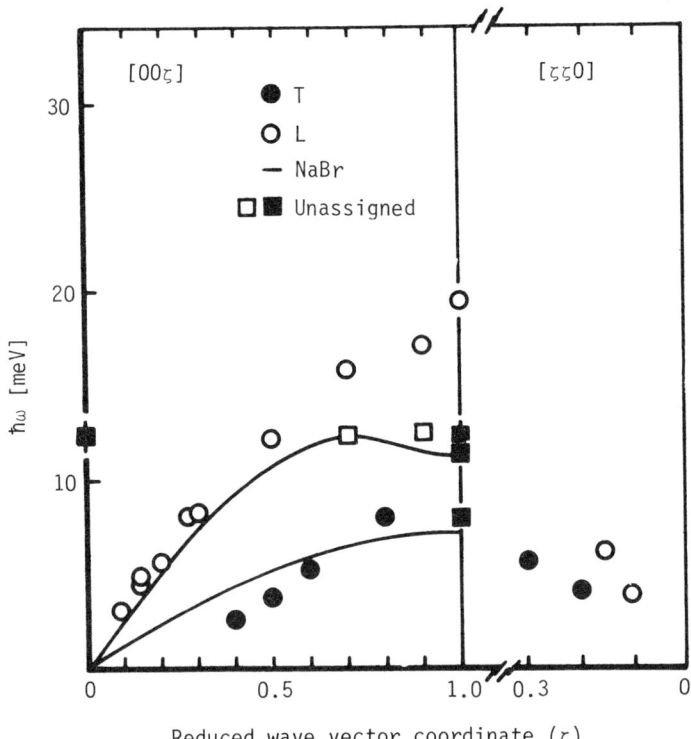

Fig. 7.8. NaCN: ω(q) [Ref. 7.21, Fig. 2], T = RT, (solid lines: dispersion curves of NaBr), Lit. [7.22,23]

KCN

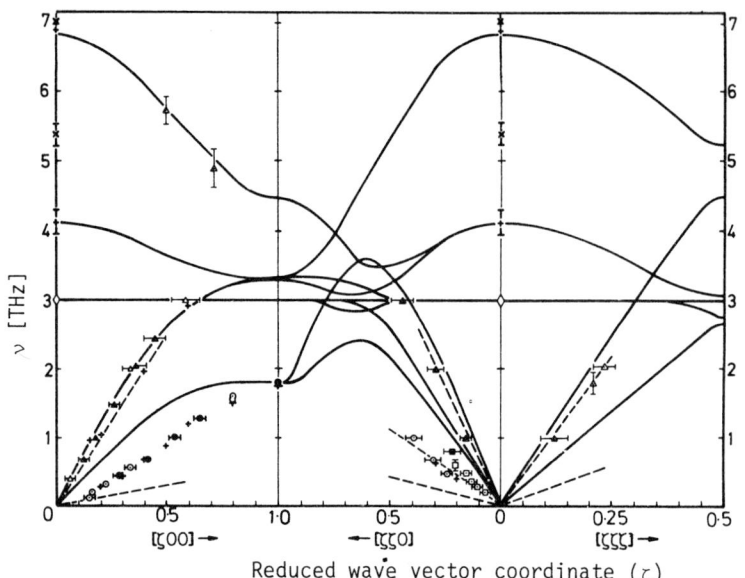

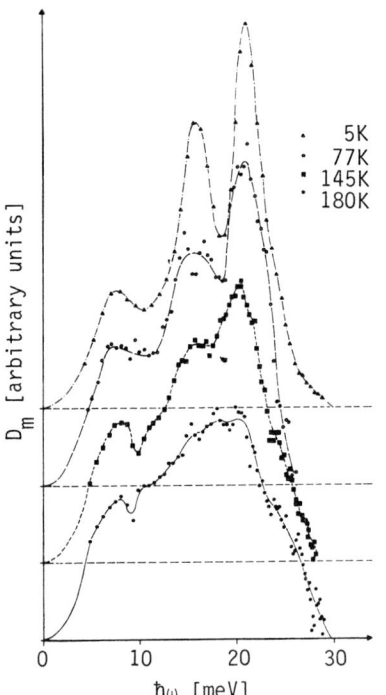

Fig. 7.9a. KCN: ω(q) [Ref. 7.24, Fig. 1], T = 215 K: O TA, □ T_2A, △ LA; T = 175 K: ⊙ TA, ▣ T_2A, △ LA, LO; ●, ▲, ▫ overlapping symbols, + [7.21], × [7.25], ◇ [7.26], full lines: 7P-SM+2P dumb-bell model [7.22], Lit. [7.21,23,27], further experimental data are given in [7.34]

Fig. 7.9b. KCN: D(ω) [Ref. 7.27, Fig. 1], T = 5 K, 77 K, 145 K, 180 K

ND$_4$I

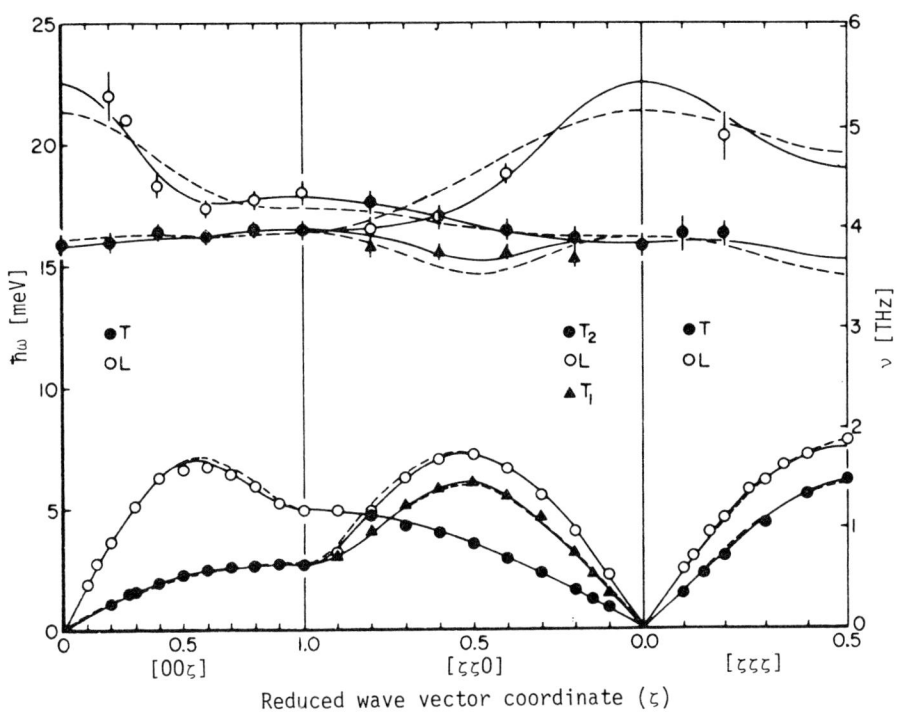

Fig. 7.10a. ND$_4$I: $\omega(q)$ [Ref.7.28, Fig.1], T = 296 K, M: 16P-BSM (full lines), 16P-SM (dashed lines), Lit. [7.29]

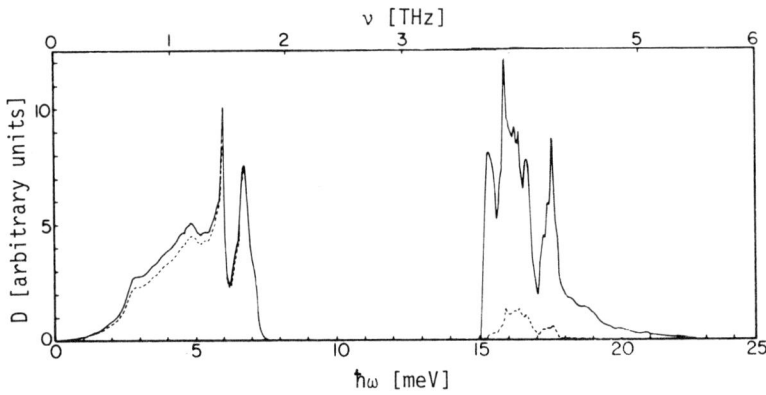

Fig. 7.10b. ND$_4$I: $D(\omega)$ [Ref. 7.28, Fig. 3], M: 16P-BSM, Lit. [7.29]

CsF

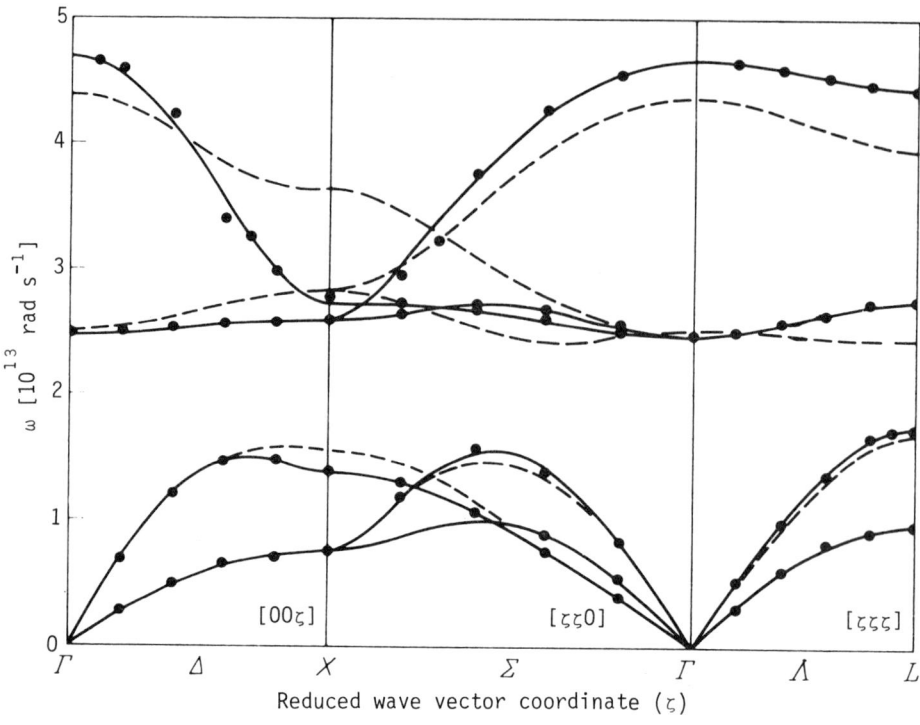

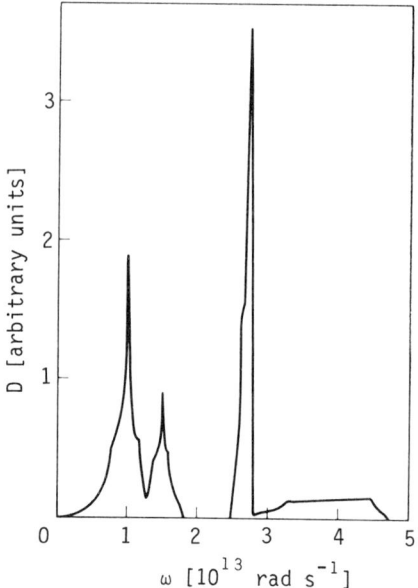

Fig. 7.11a. CsF: ω(q) [Ref. 7.30, Fig. 1], T = 80K, M: 11P-SM (full lines), 5P-FCM (dashed lines), Lit. [7.31-33]

Fig. 7.11b. CsF: D(ω) [Ref. 7.30, Fig. 3], M: 11P-SM, Lit. [7.31]

8. Cesium Chloride Structure Crystals

Crystals with this structure exhibit generally a complex dielectric behavior caused by the high polarizability of the constituent ions. Under pressure, several of them undergo phase transitions into the rock salt structure. Strong anharmonic effects are connected with these properties. The ammonium-halides are closely related to those with rock salt structures described in Chap.7.

Crystal	Figures showing	
	Dispersion curve $\omega(\underline{q})$	Density of states $D(\omega)$
CsCl	8.1a	8.1b
CsBr	8.2a	8.2b
CsI	8.3a	8.3b
NH_4Cl	8.4a	8.4b
ND_4Cl	8.5a	8.5b
ND_4Br	8.6	
TlCl	8.7	
TlBr	8.8a	8.8b

CsCl

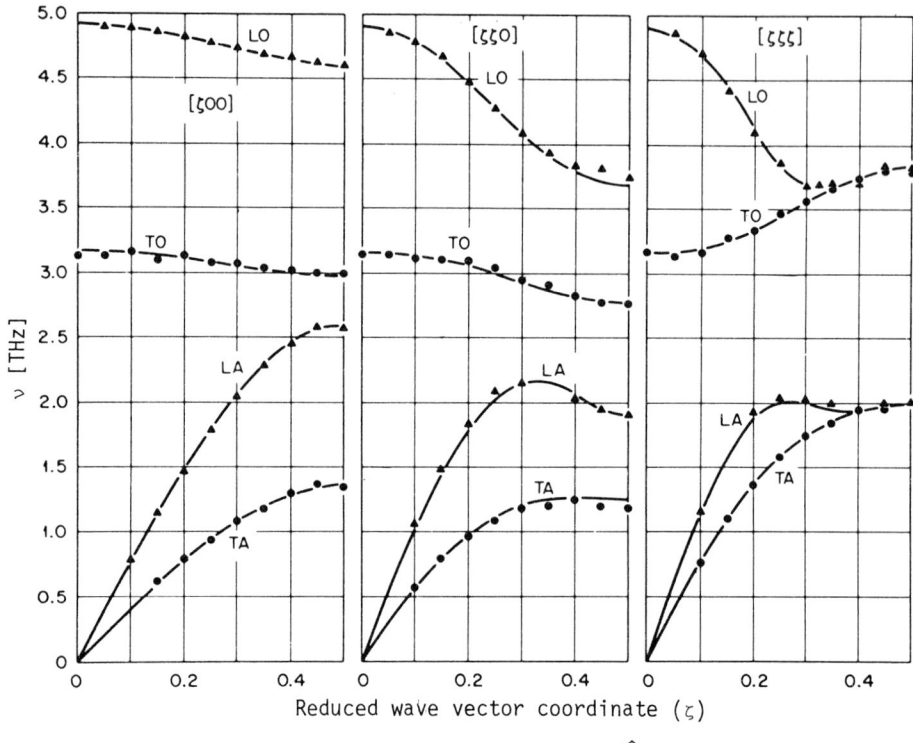

Fig. 8.1a. CsCl: $\omega(\underline{q})$ [Ref. 8.1, Fig. 1], T = 78 K, M: 11P-SM, Lit. [8.2-4,22]

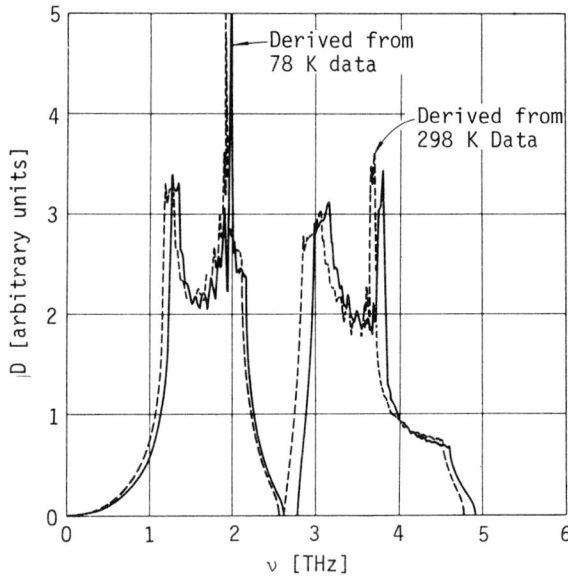

Fig. 8.1b. CsCl: $D(\omega)$ [Ref. 8.1, Fig. 2], M: 11P-SM, T = 78 K (full lines), T = 298 K (dashed lines)

CsBr

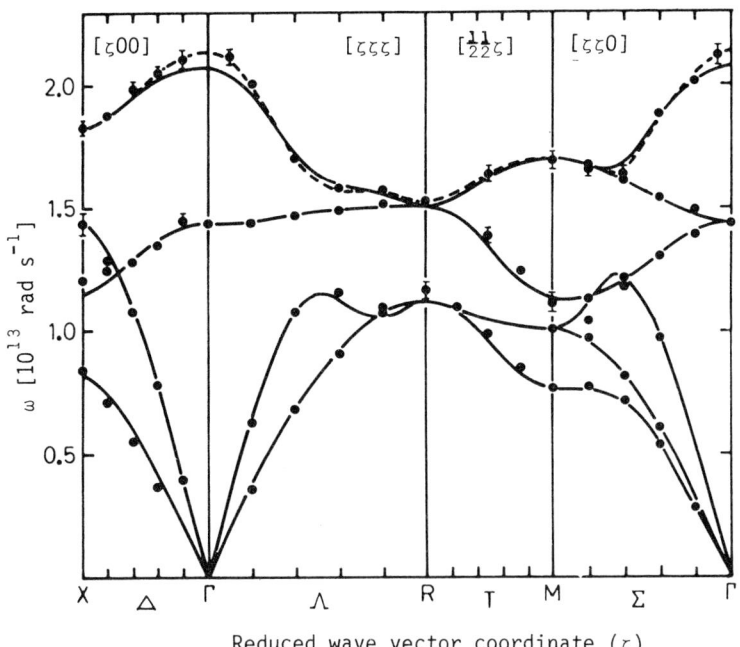

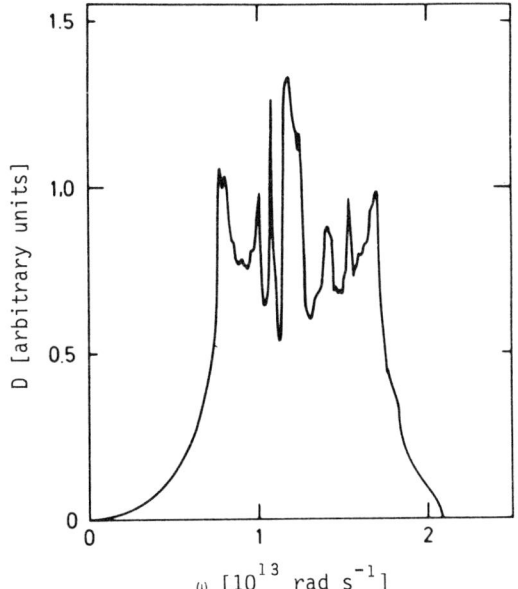

Fig. 8.2a. CsBr: ω(q) [Ref. 8.5, Fig. 1], T = 80 K, M: 11P-SM (full lines), 11P-BSM (dashed lines), Lit. [8.2,3,6-9,22]

Fig. 8.2b. CsBr: D(ω) [Ref. 8.5, Fig. 2], T = 80 K, M: 11P-SM, Lit. [8.9]

CsI

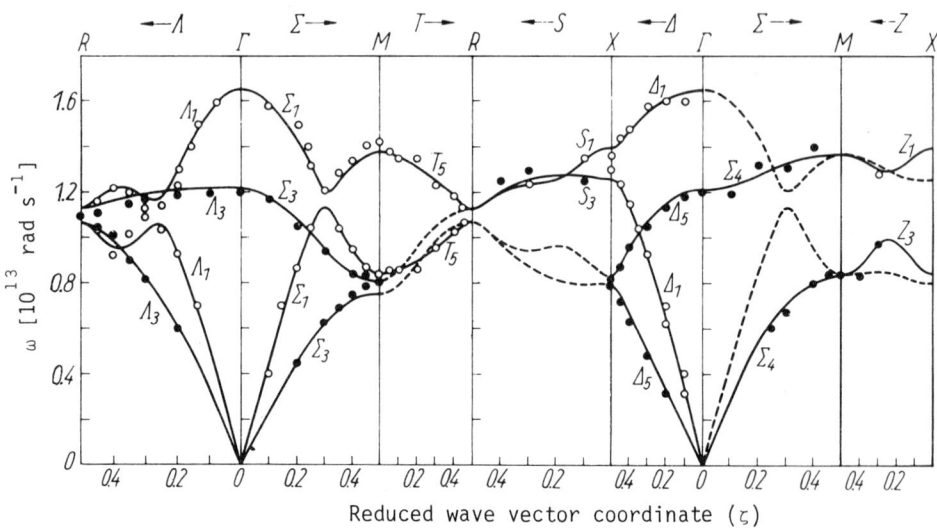

Fig. 8.3a. CsI: $\omega(q)$ [Ref. 8.10, Fig. 2], T = 293 K, M: 14P-SM, Lit. [8.2,3,8,11,12,22]

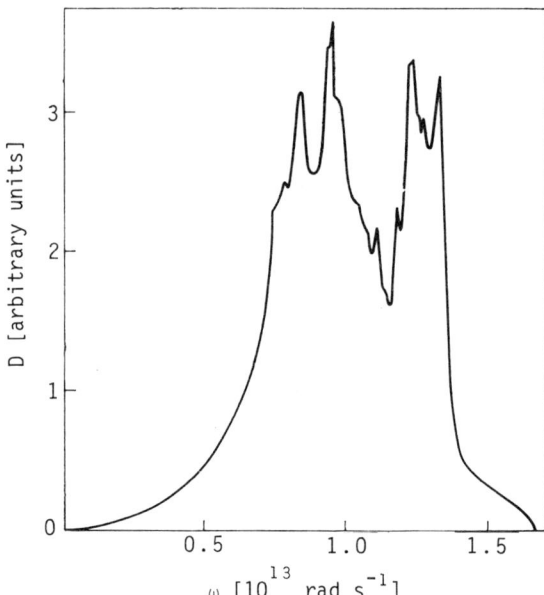

Fig. 8.3b. CsI: $D(\omega)$ [Ref. 8.10, Fig. 3], T = 293 K, M: 14P-SM

NH_4Cl

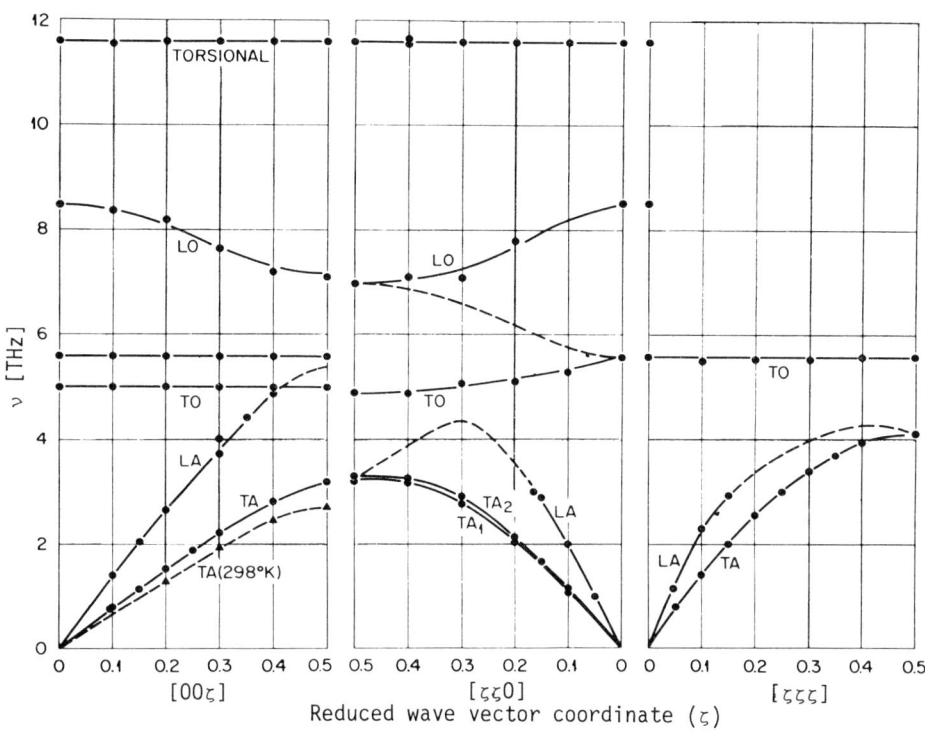

Fig. 8.4a. NH_4Cl: $\omega(\underline{q})$ [Ref. 8.13, Fig. 16], T = 78 K, M: -, Lit. [8.14-17]

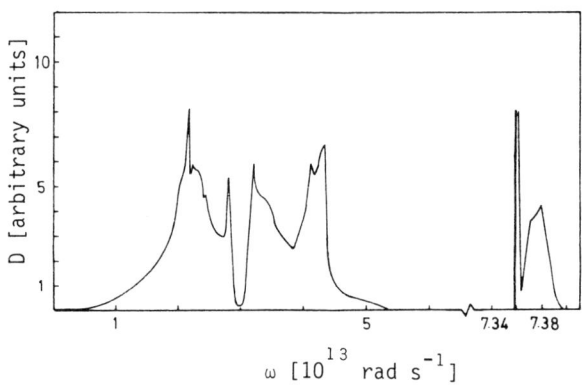

Fig. 8.4b. NH_4Cl: $D(\omega)$ [Ref. 8.14, Fig. 3], T = 78 K, M: 5P-FCM, Lit. [8.17]

ND_4Cl

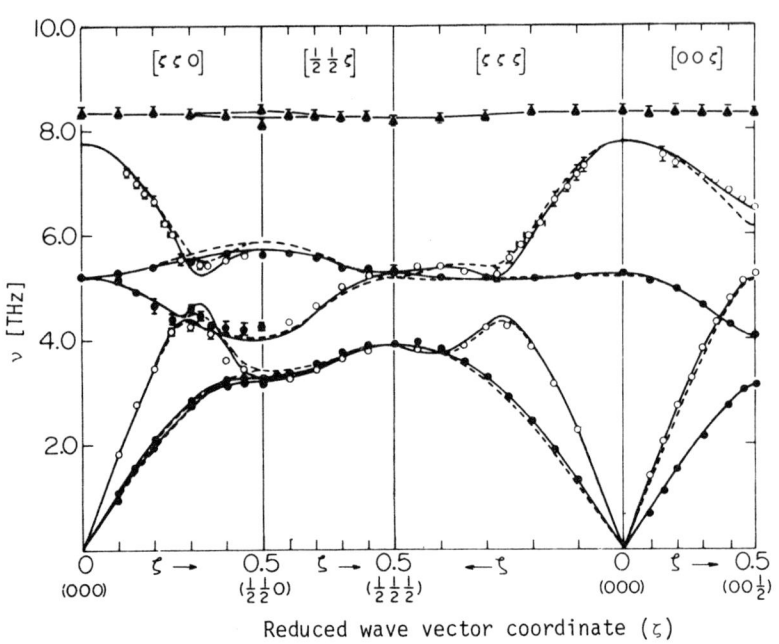

Fig. 8.5a. ND_4Cl: $\omega(\underline{q})$ [Ref.8.18, Fig.1], T = 85 K, M: 13P-SM (solid lines), 11P-FCM (dashed lines), Lit. [8.14,16,17,19-21]

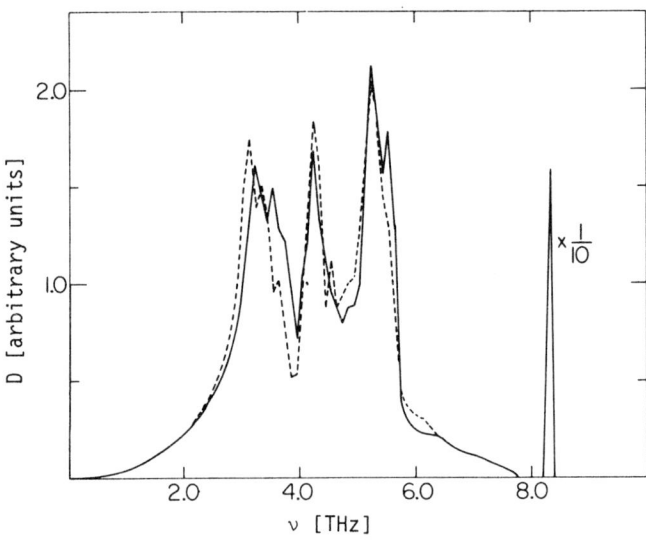

Fig. 8.5b. ND_4Cl: $D(\omega)$ [Ref. 8.18, Fig. 2], T = 85 K, M: 13P-SM (full lines), 16P-SM [18.16] (dashed lines), Lit. [8.16]

ND₄Br

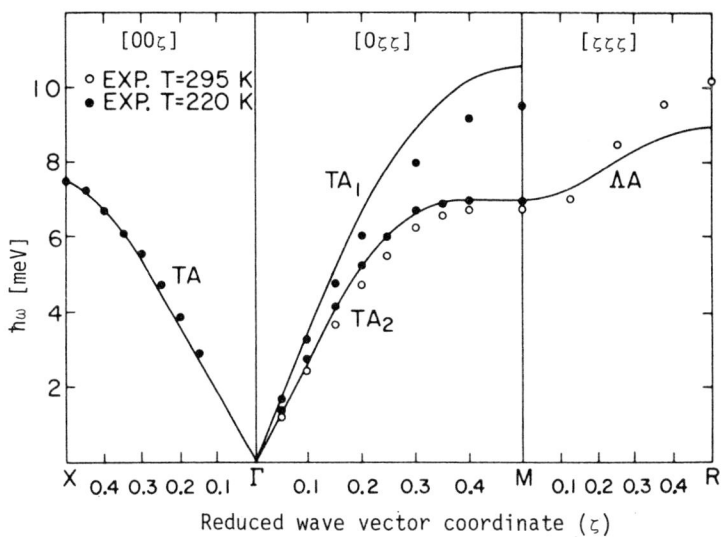

Fig. 8.6. ND$_4$Br: $\omega(\underline{q})$ [Ref. 8.23, Fig. 2], T = 220 K, 295 K, M: RIM

TlCl

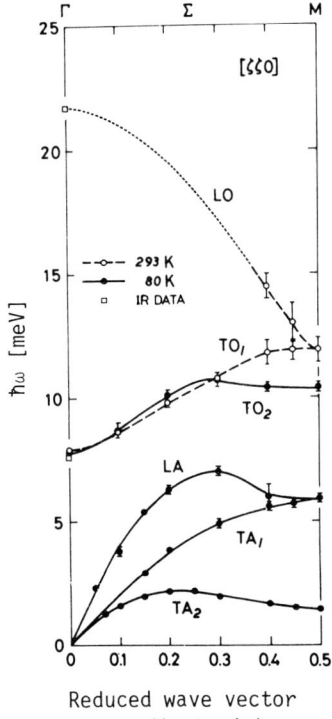

Fig. 8.7. TlCl: $\omega(\underline{q})$ [Ref. 8.24, Fig. 2]
T = 293 K, 80 K

TlBr

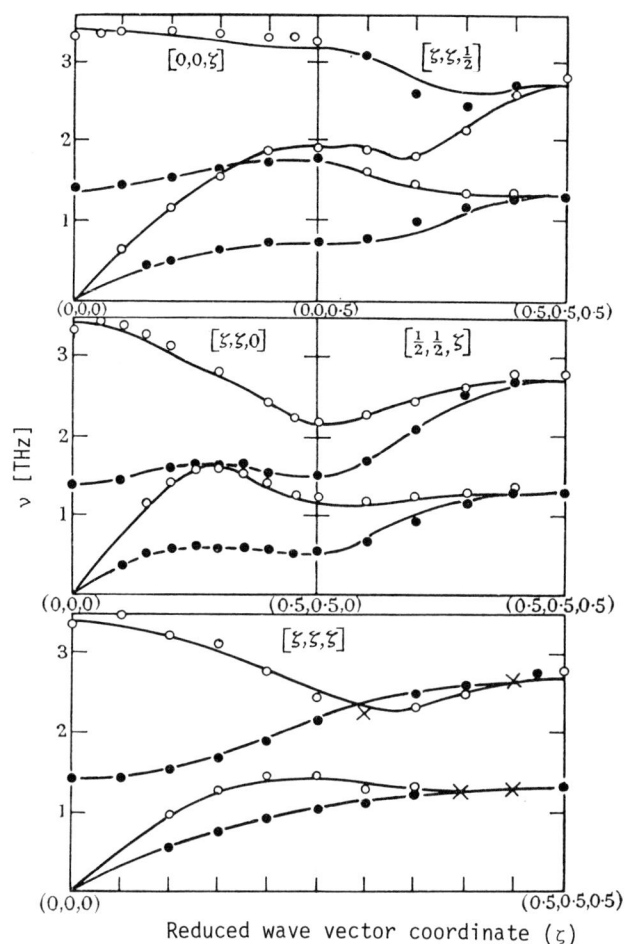

Fig. 8.8a. TlBr: $\omega(\underline{q})$ [Ref. 8.25, Fig. 2], T = 100 K, M: 14P-SM, Lit. [8.26]

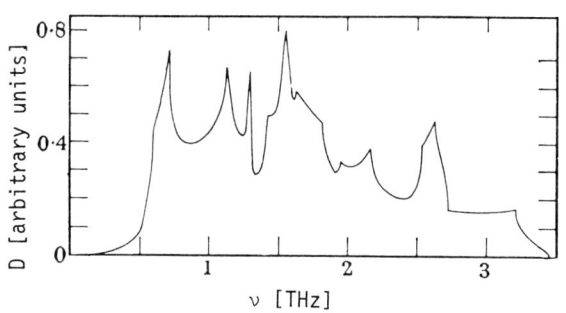

Fig. 8.8b. TlBr: $D(\omega)$ [Ref. 8.25, Fig. 3a], T = 100 K, M: 14P-SM

9. Diamond Structure Crystals

The covalency of the elemental semiconductors may be described in different ways in the different phonon models. In the shell model, COCHRAN [9.25] found that interionic core-shell interactions are important for the description of the flat TA branches in Si, Ge, and α-Sn. An alternative approach stems from the molecular picture using valence field forces [9.10]. The most pictorial way of description is the bond charge model (BCM) which in its general form [9.12] is a very powerful method to understand the mechanical properties of these crystals, although the dielectric properties are not well described.

The following systems are treated:

Crystal	Figures showing	
	Dispersion curve $\omega(\underline{q})$	Density of states $D(\omega)$
Diamond	9.1a	9.1b
Si	9.2a	9.2b
Ge	9.3a	9.3b
α-Sn	9.4a	9.4b

Diamond

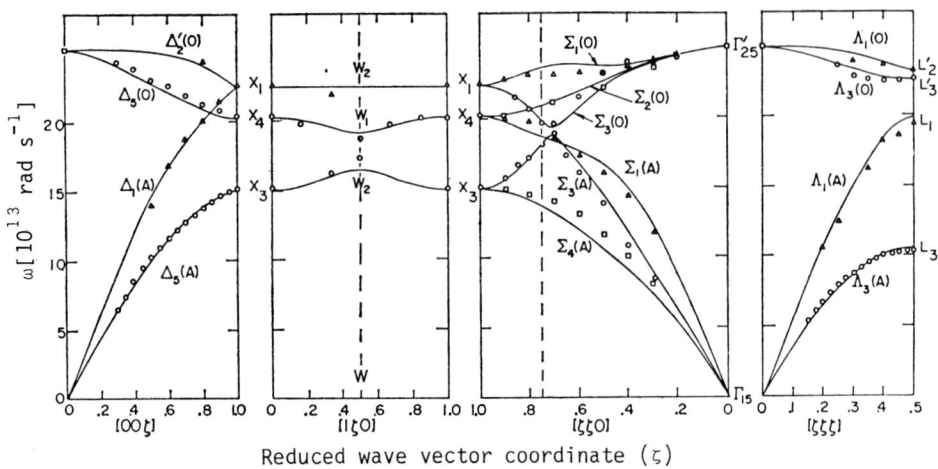

Fig. 9.1a. Diamond: $\omega(\underline{q})$ [Ref. 9.1, Fig. 1], T = 296 K, M: 10P-SM, Lit. [9.2-12]

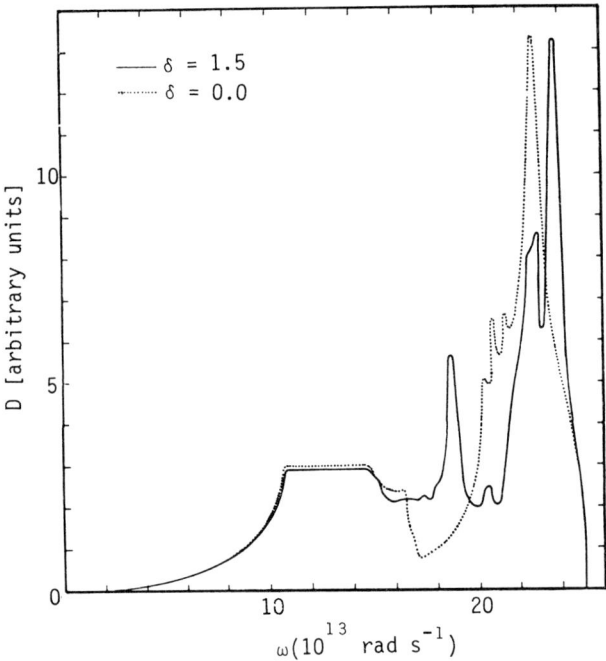

Fig. 9.1b. Diamond: $D(\omega)$ [Ref. 9.2, Fig. 3], T = 296 K, M: 10P-SM, Lit. [9.3,10,12]

Si

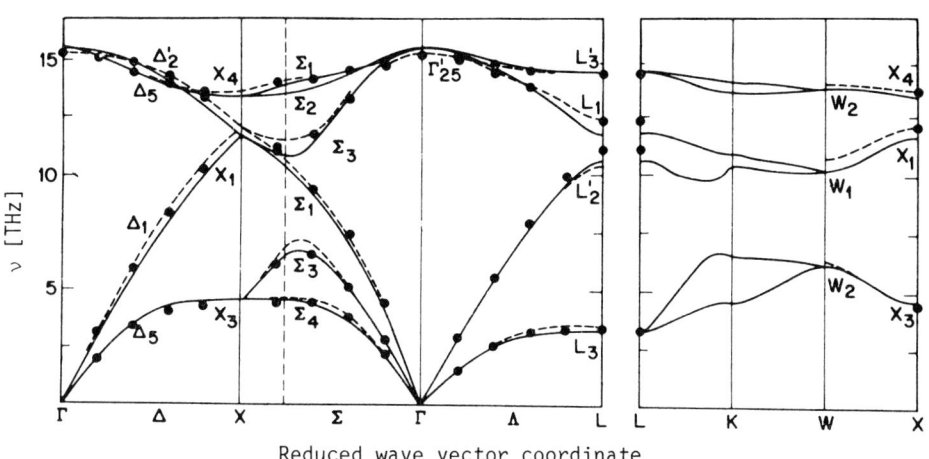

Fig. 9.2a. Si: ω(q) [Ref. 9.12, Fig. 3], T = RT, M: 4P-BCM (full lines), 6P-VFFM [9.10] (dashed lines) measurements [9.13,14], Lit. [9.3,7-19]

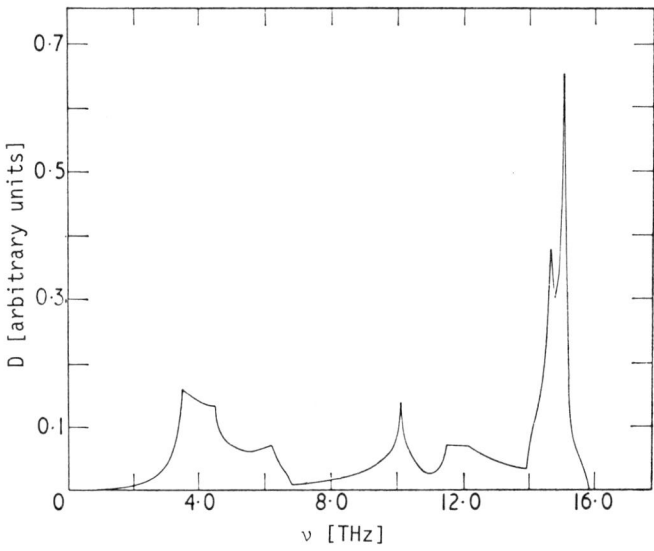

Fig. 9.2b. Si: D(ω) [Ref. 9.3, Fig. 2b], T = RT, M: 4P-BCM, Lit. [9.10,12]

Ge

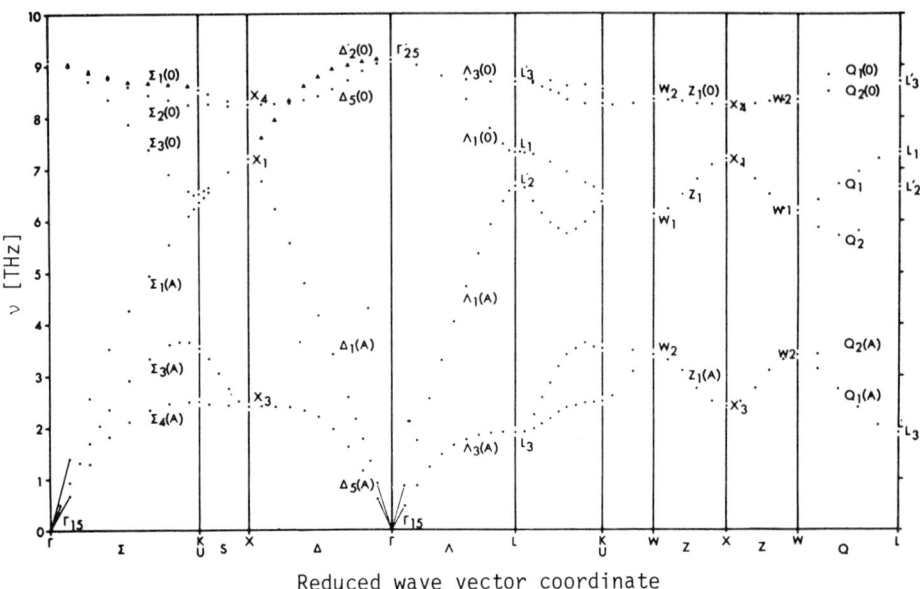

Fig. 9.3a. Ge: $\omega(\underline{q})$ [Ref. 9.20, Fig. 2], T = 80 K, Lit. [9.3,7-12,18,21-31]

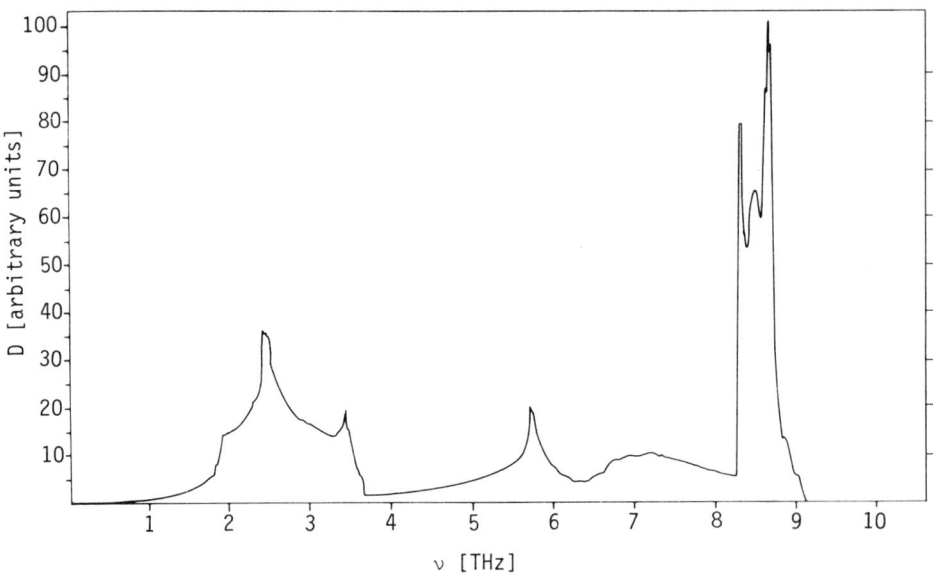

Fig. 9.3b. Ge: $D(\omega)$ [Ref. 9.21, Fig. 3], T = 80 K, Lit. [9.3,10,12]

α-Sn

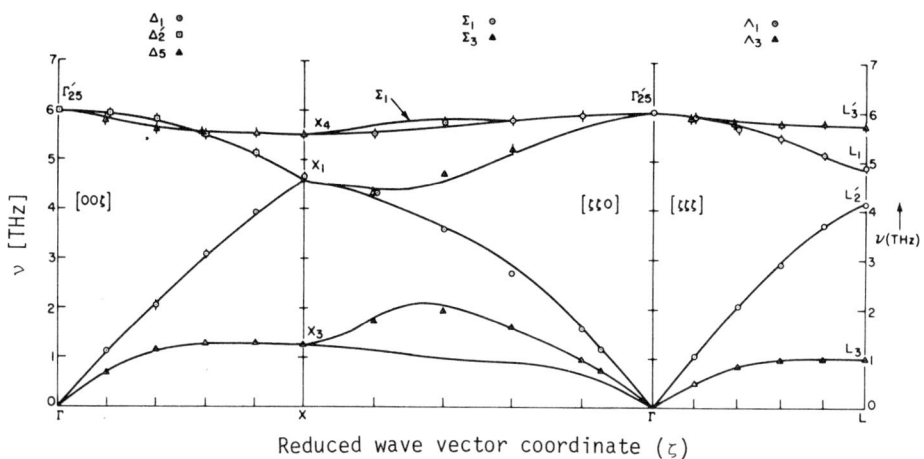

Fig. 9.4a. α-Sn: ω(q) [Ref. 9.32, Fig. 1], T = 90 K, M: 11P-SM, Lit. [9.10,12,26,30,33-35]

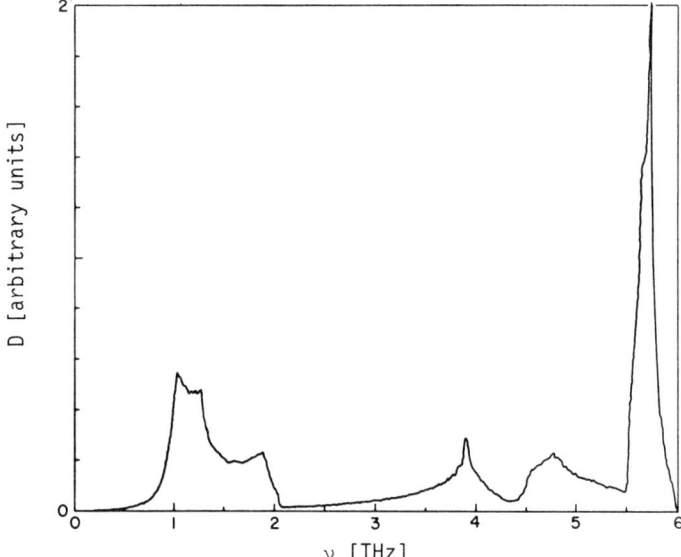

Fig. 9.4b. α-Sn: D(ω) [Ref. 9.33, Fig. 2a], T = 90 K, M: 11P-SM, Lit. [9.10,12]

10. Zinc-Blende Structure Crystals

Crystals with this structure exhibit a complex mixture of covalent and ionic properties together with a generally high polarizability of the lattice ions. In crystals which are close to a phase transition to a more ionic structure (such as CuCl), strong anharmonic effects were observed [10.37,38]. The most successful model seems, at present, a combination of valence, ionic, and polarization forces including the overlap polarizability [10.11]. Another interesting is the asymmetric version of the bond charge model [10.12].

The following systems are treated:

Crystal	Figures showing		Crystal	Figures showing	
	Dispersion curve $\omega(\underline{q})$	Density of states $D(\omega)$		Dispersion curve $\omega(\underline{q})$	Density of states $D(\omega)$
SiC	10.1a	10.1b	ZnS	10.9a	10.9b
AlSb	10.2a	10.2b	ZnSe	10.10a	10.10b
GaP	10.3a	10.3b	ZnTe	10.11a	10.11b
GaAs	10.4a	10.4b	CdTe	10.12a	10.12b
GaSb	10.5a	10.5b	CuCl	10.13a	10.13b
InP	10.6a	10.6b	CuBr	10.14a	10.14b
InAs	10.7		CuI	10.15a	10.15b
InSb	10.8a	10.8b			

SiC

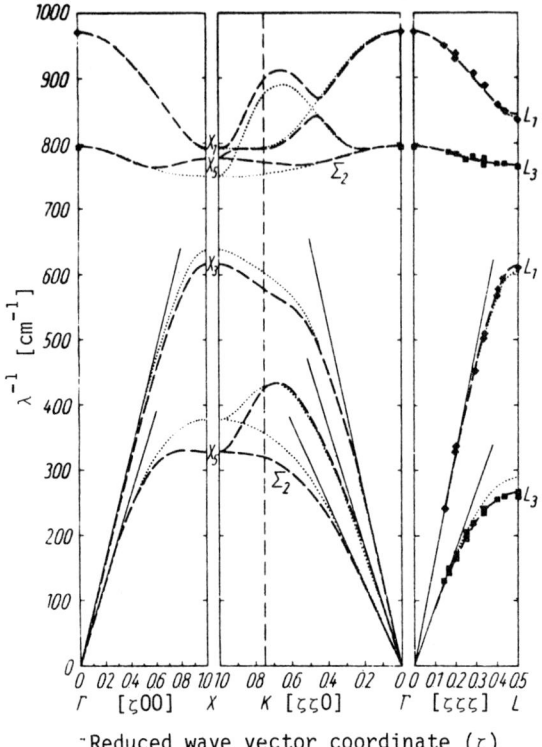

Fig. 10.1a. SiC: $\omega(\underline{q})$ [Ref. 10.1, Fig. 2h] M: 15P-DBM (full lines), 13P-DBM (dashed lines), 9P-DBM (dotted lines), experimental data [10.2], Lit. [10.3-6]

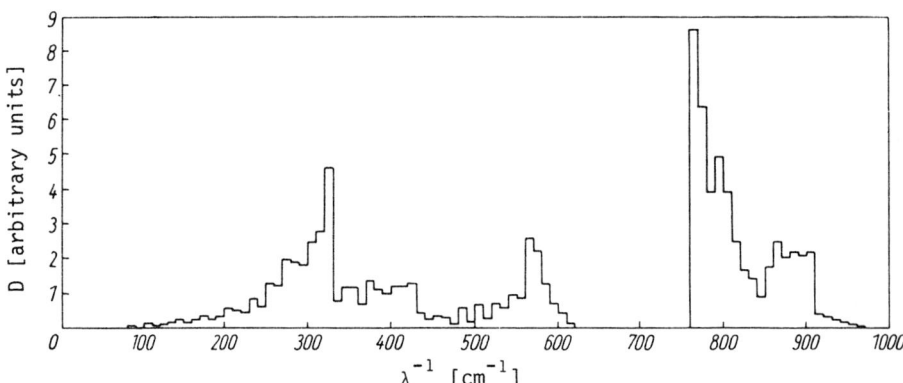

Fig. 10.1b. SiC: $D(\omega)$ [Ref. 10.1, Fig. 3h], M: 15P-DBM, Lit. [10.3,4]

AlSb

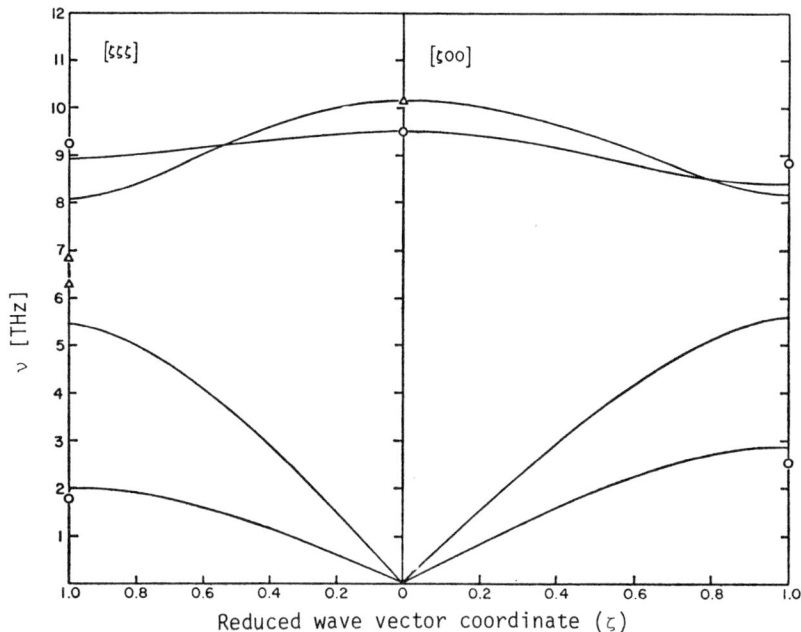

Fig. 10.2a. AlSb: $\omega(\underline{q})$ [Ref. 10.7, Fig. 2], M: 7P-RIM

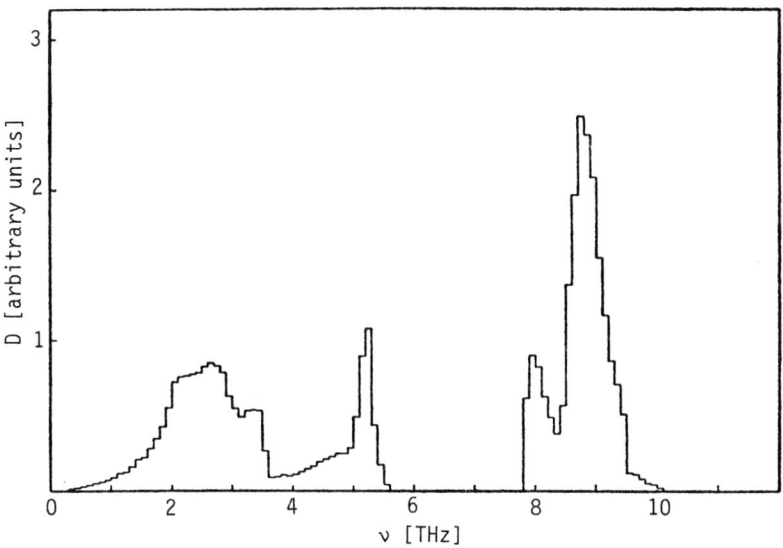

Fig. 10.2b. AlSb: $D(\omega)$ [Ref. 10.7, Fig. 3], M: 7P-RIM

GaP

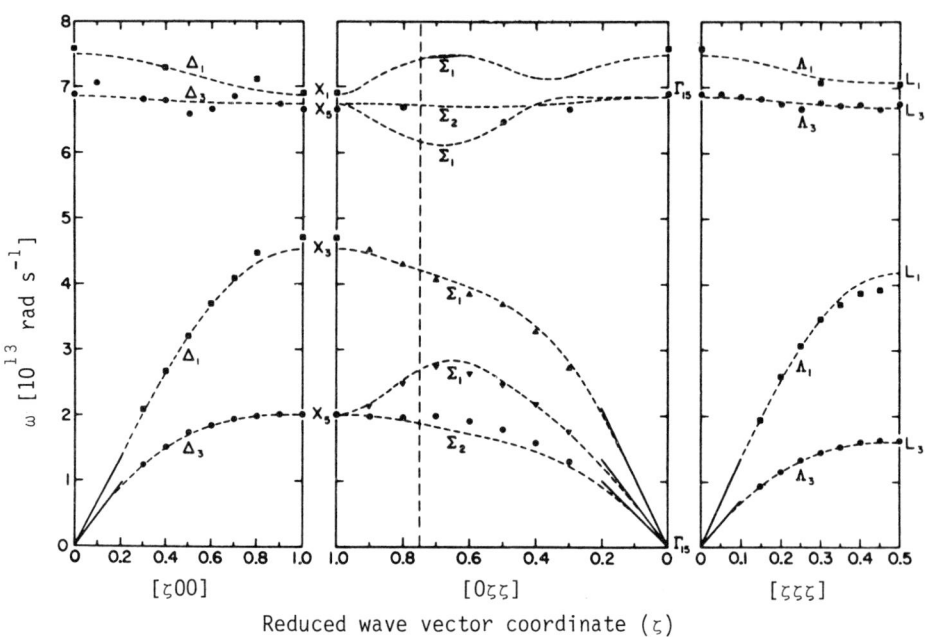

Fig. 10.3a. GaP: ω(q) [Ref. 10.8, Fig. 3], T = RT, M: 14P-SM, Lit. [10.9-12]

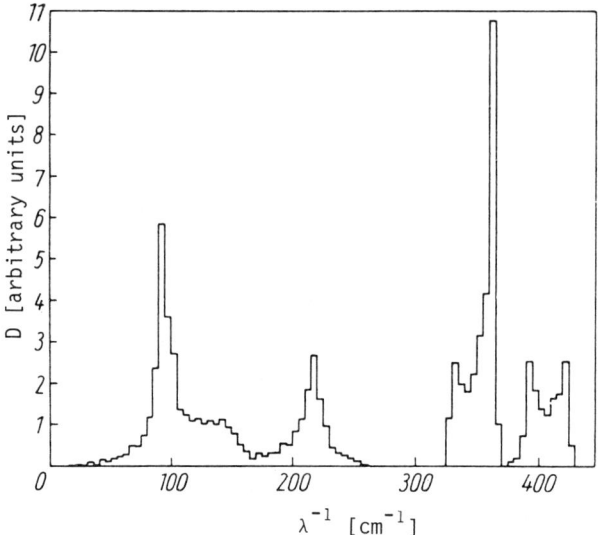

Fig. 10.3b. GaP: D(ω) [Ref. 10.10, Fig. 3c], M: 15P-DBM, Lit. [10.9]

GaAs

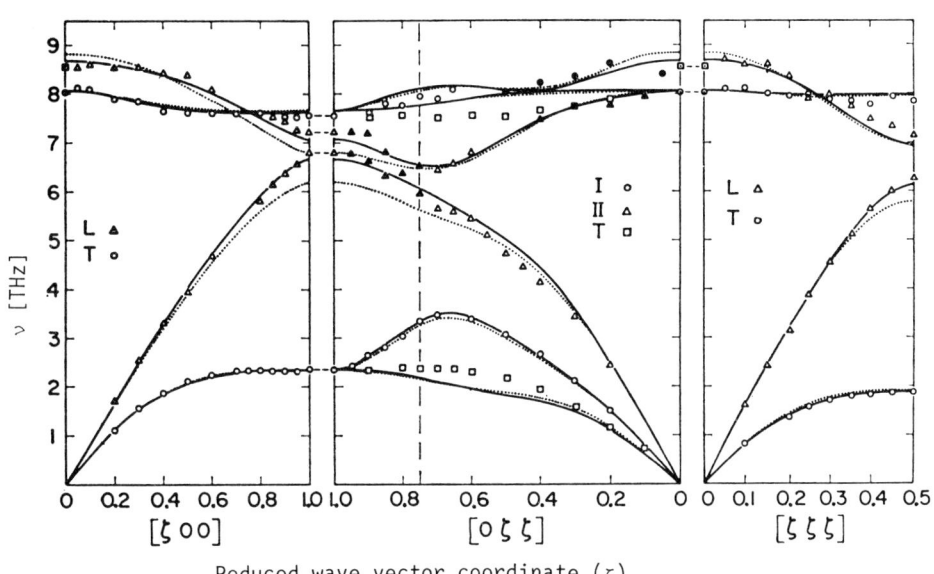

Fig. 10.4a. GaAs: ω(q) [Ref. 10.13, Fig. 1], T = 296 K, M: 14P-SM (full lines), 12P-SM (dashed lines), Lit. [10.7,10-13,15]

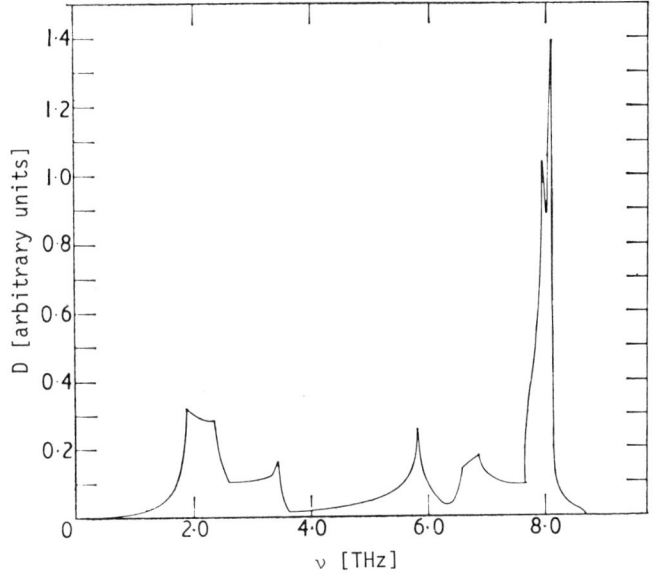

Fig. 10.4b. GaAs: D(ω) [Ref. 10.14, Fig. 2d], T = 296 K, M: 14P-SM, Lit. [10.7,10,13,14]

GaSb

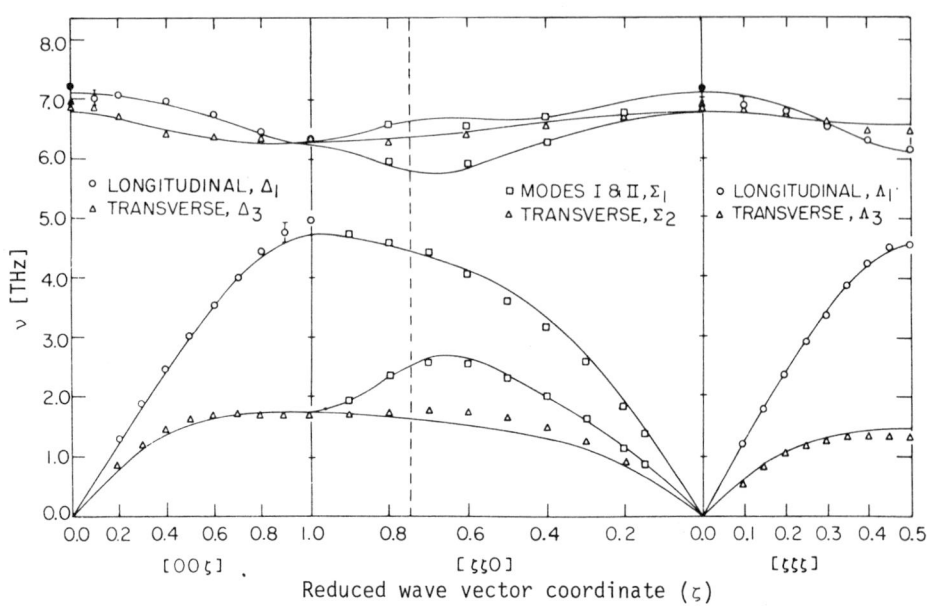

Fig. 10.5a. GaSb: $\omega(\underline{q})$ [Ref. 10.16, Fig. 2], T = RT, M: 14P-SM, Lit. [10.11,12,17]

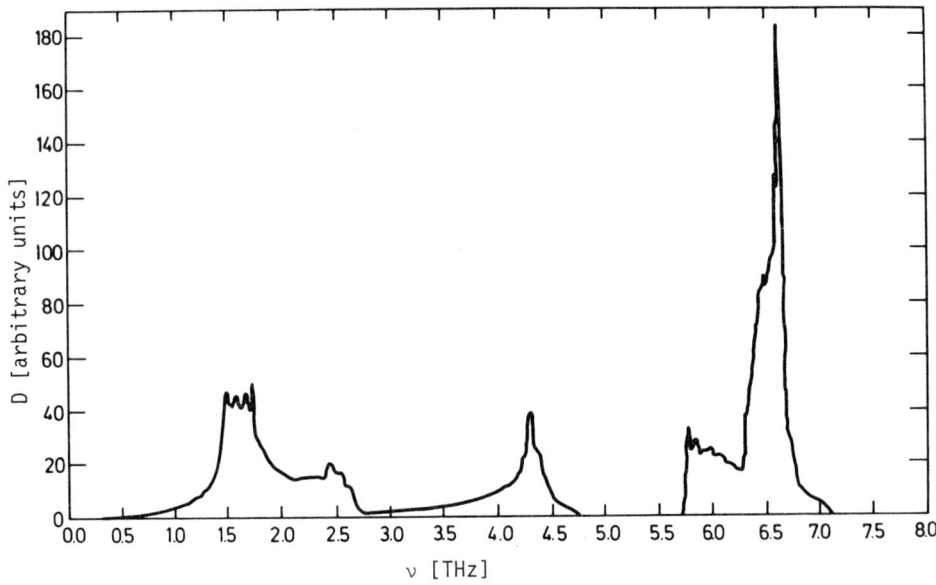

Fig. 10.5b. GaSb: $D(\omega)$ [Ref. 10.16, Fig. 4], T = RT, M: 14P-SM

InP

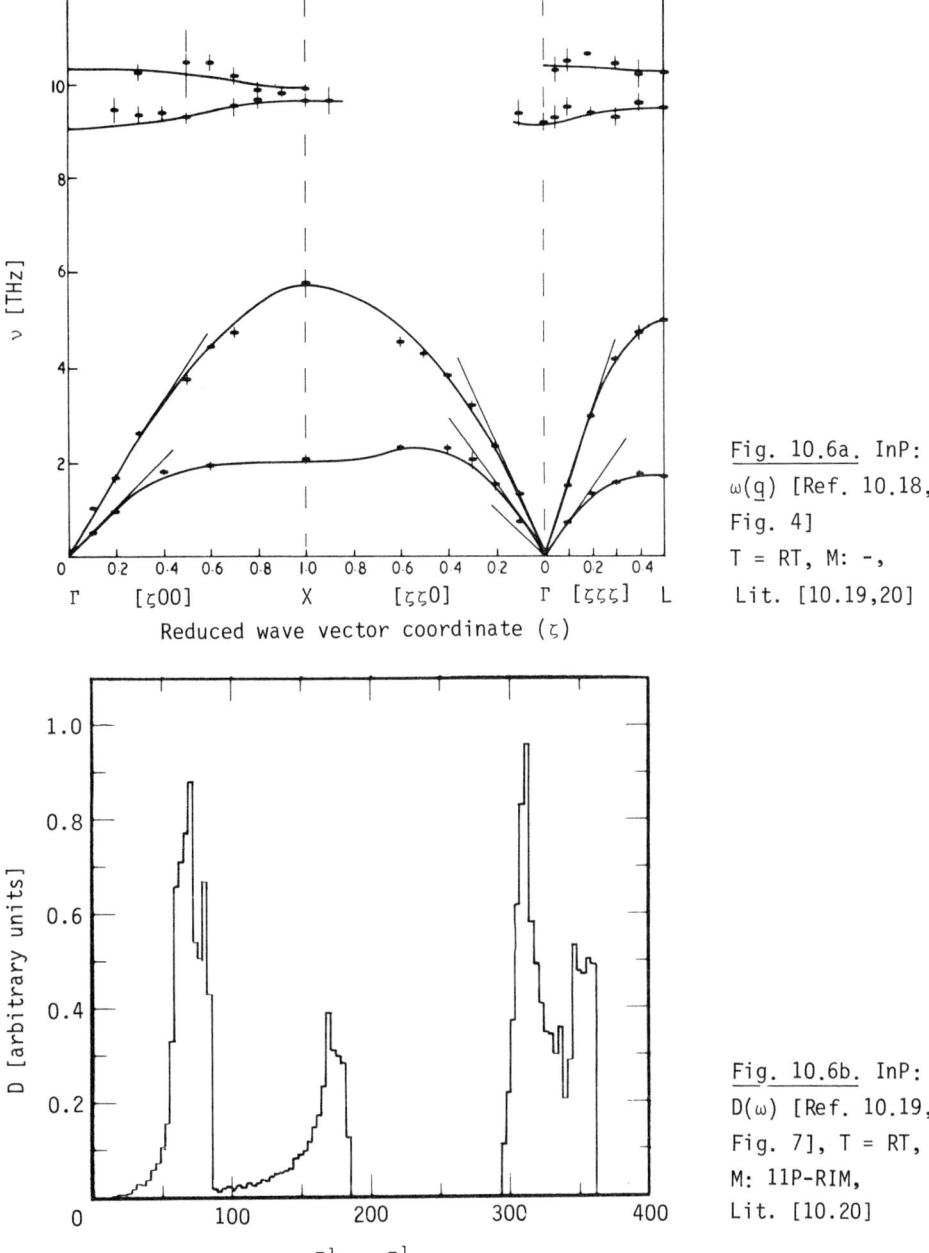

Fig. 10.6a. InP: $\omega(\underline{q})$ [Ref. 10.18, Fig. 4] T = RT, M: -, Lit. [10.19,20]

Fig. 10.6b. InP: $D(\omega)$ [Ref. 10.19, Fig. 7], T = RT, M: 11P-RIM, Lit. [10.20]

InAs

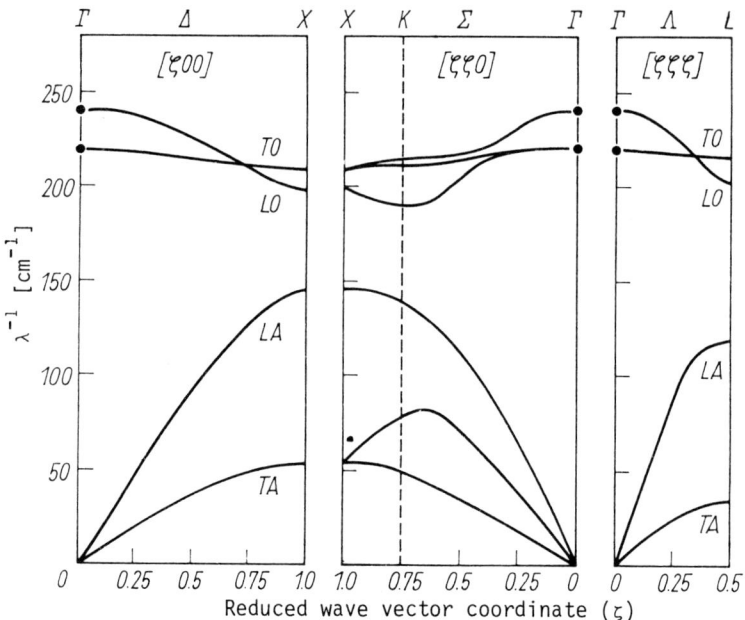

Fig. 10.7. InAs: $\omega_c(\underline{q})$ [Ref. 10.30, Fig. 6], M: 7P-RIM, Lit. [10.20]

InSb

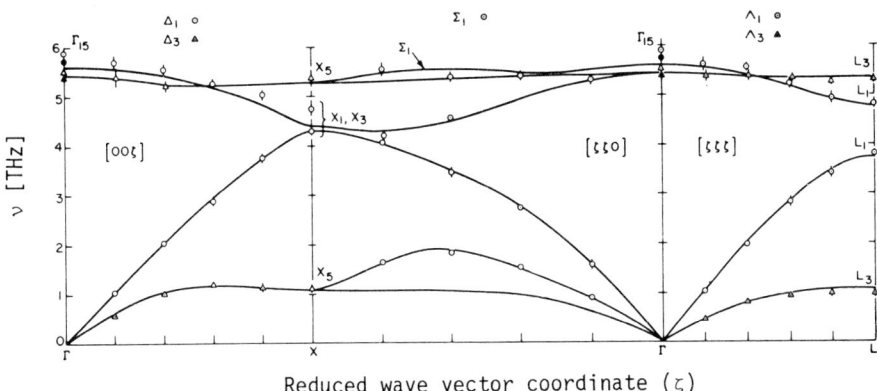

Fig. 10.8a. InSb: ω(q) [Ref. 10.21, Fig. 2], T = RT, M: 14P-SM, Lit. [10.10,12,20,22÷26]

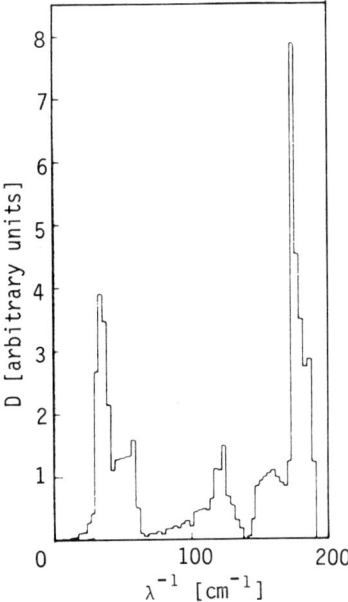

Fig. 10.8b. InSb: D(ω) [Ref. 10.10., Fig. 3e], M: 15P-DBM, Lit. [10.10,20,22]

ZnS

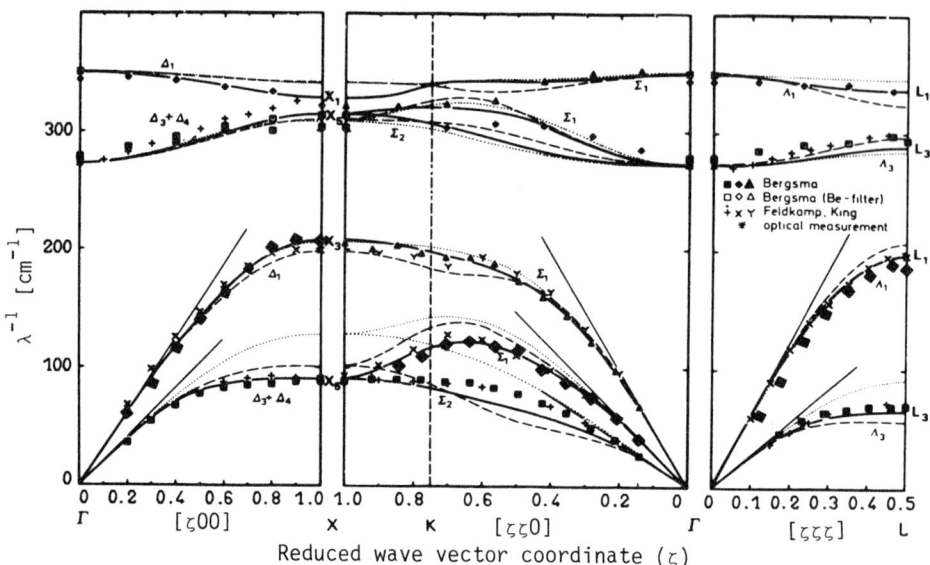

Fig. 10.9a. ZnS: ω(q) [Ref. 10.25, Fig. 1a], T = RT, M: 15P-DBM (full lines), 13P-DBM (dashed lines), 9P-DBM (dotted lines), measurements [10.27,28], Lit. [10.9,10,25,27-29,31,32]

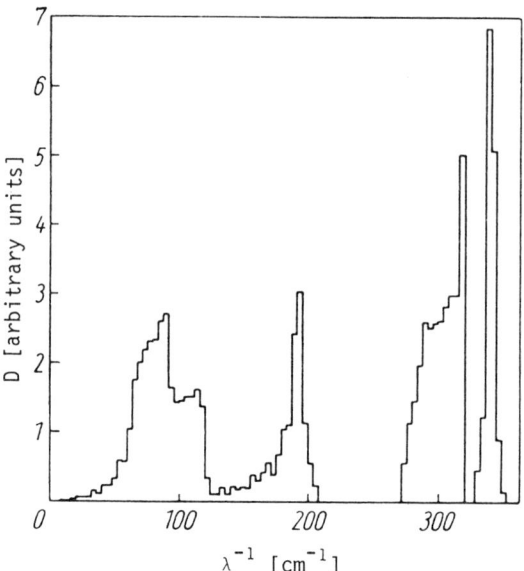

Fig. 10.9b. ZnS: D(ω) [Ref. 10.10, Fig. 3a], M: 15P-DBM, Lit. [10.9,10,31,32]

ZnSe

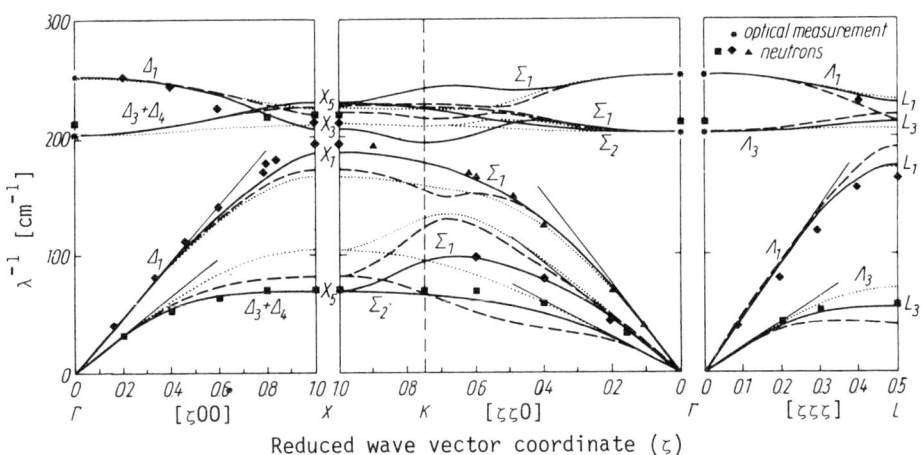

Fig. 10.10a. ZnSe: ω(q) [Ref. 10.10, Fig. 2b], T = RT, M: 15P-DBM (full lines) 13P-DBM (dashed lines), 9P-DBM (dotted lines), measurements [10.33], Lit. [10.11,23,33]

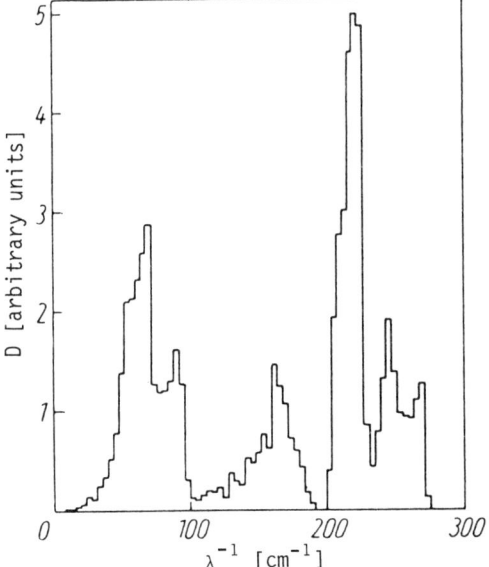

Fig. 10.10b. ZnSe: D(ω) [Ref. 10.10, Fig. 3b], M: 15P-DBM, Lit. [10.33]

ZnTe

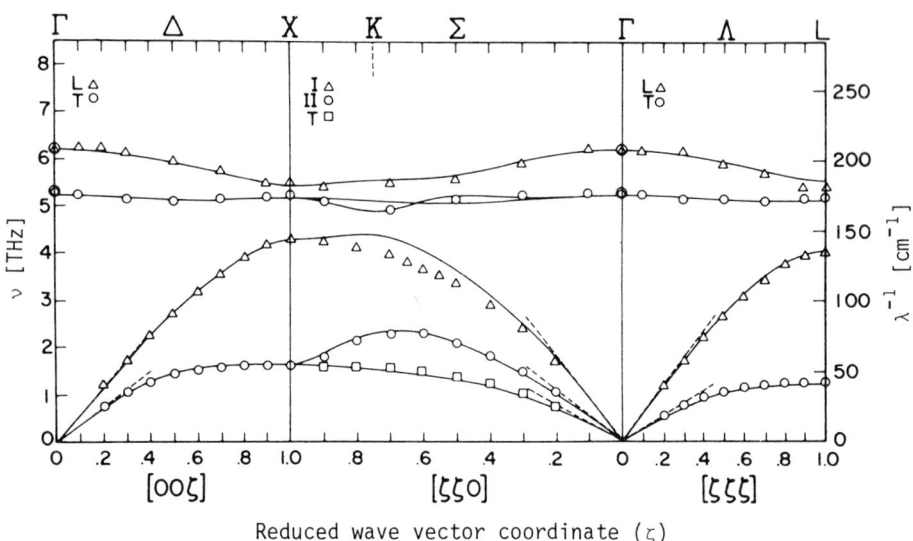

Fig. 10.11a. ZnTe: $\omega(\underline{q})$ [Ref. 10.32, Fig. 1], T = RT, M: 10P-VSM, Lit. [10.34]

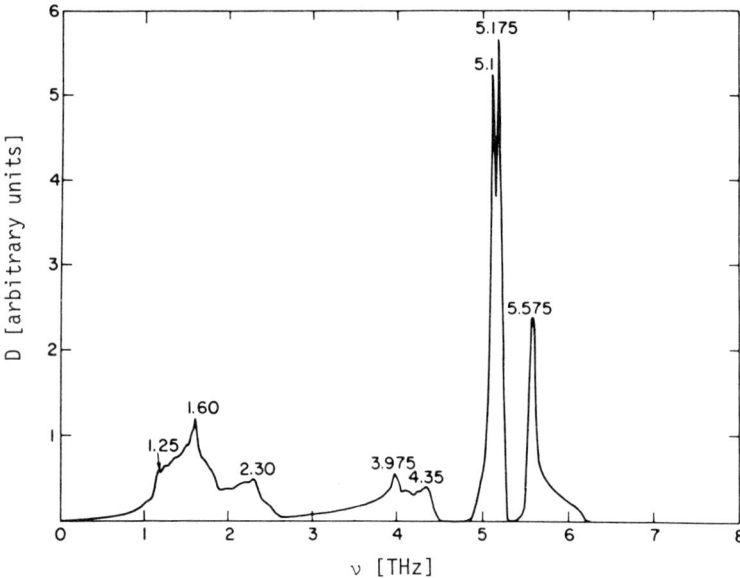

Fig. 10.11b. ZnTe: $D(\omega)$ [Ref. 10.32, Fig. 2], T = RT, M: 10P-VSM, Lit. [10.34]

CdTe

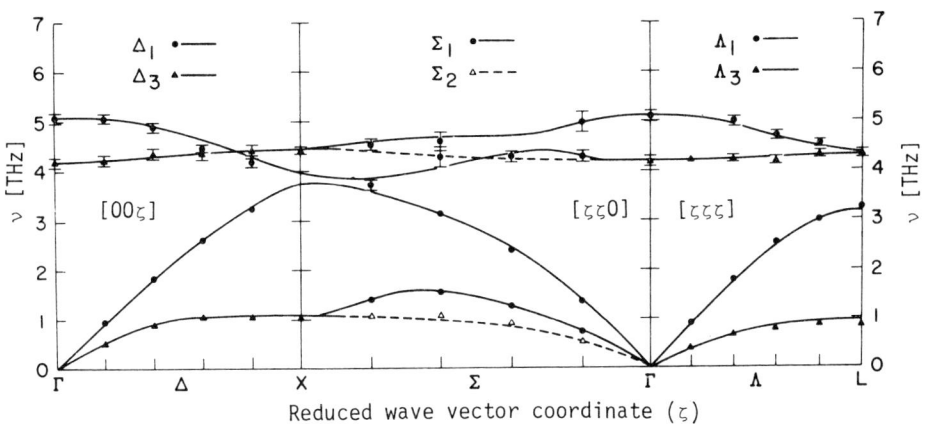

Fig. 10.12a. CdTe: $\omega(\underline{q})$ [Ref. 10.35, Fig. 1], T = 300 K, M: 14P-SM, Lit. [10.34,36]

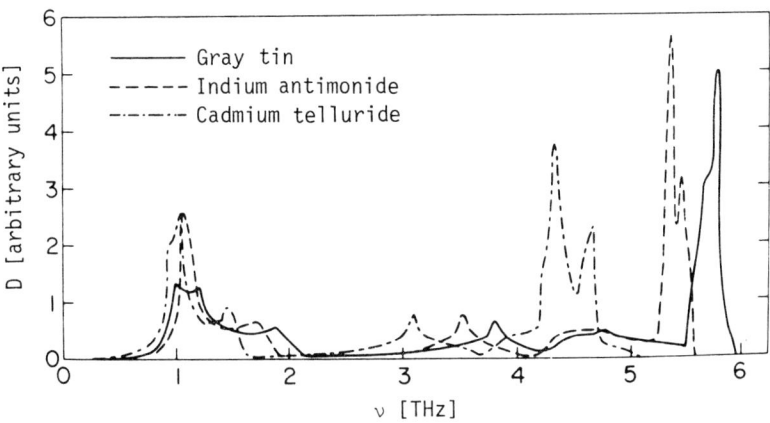

Fig. 10.12b. CdTe: $D(\omega)$ [Ref. 10.35, Fig. 3], M: 14P-SM, Lit. [10.34]

CuCl

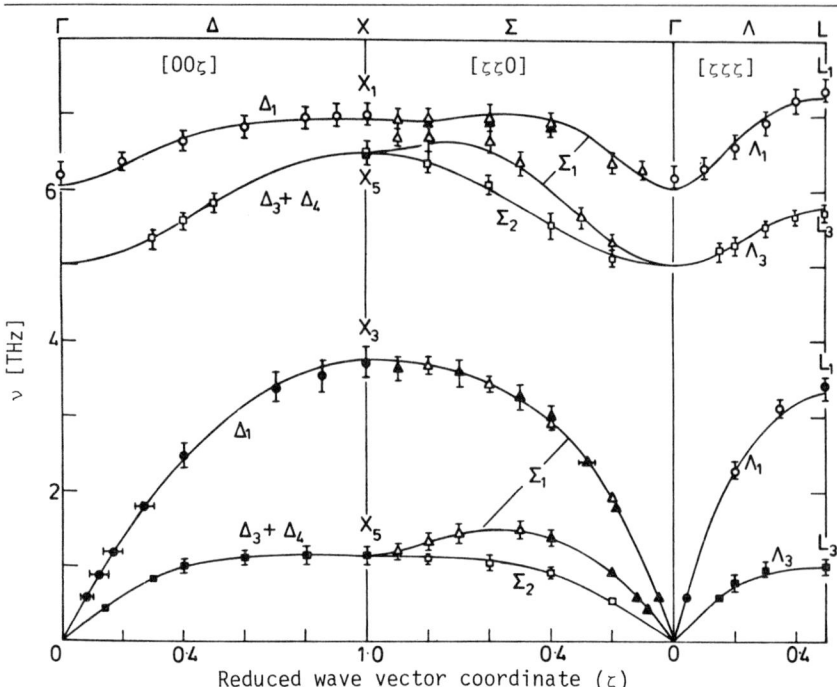

Fig. 10.13a. CuCl: ω(q) [Ref. 10.37, Fig. 1], T = 4.2 K, M: 14P-SM, Lit. [10.10,38,39]

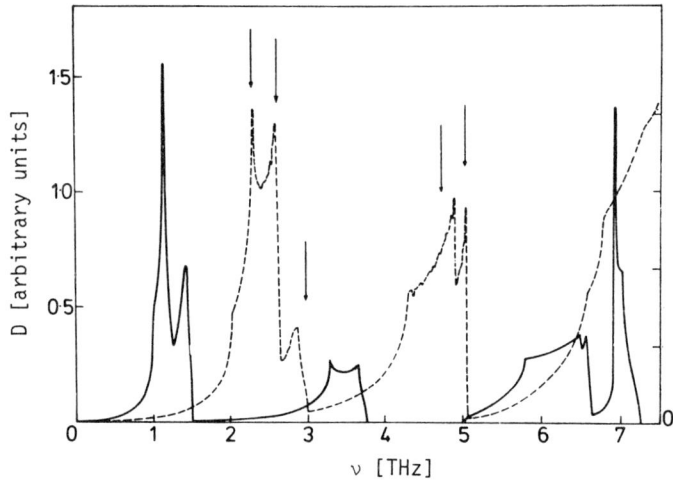

Fig. 10.13b. CuCl: D(ω) [Ref. 10.37, Fig. 5], one phonon denisty (full lines), combined density of states (dashed lines), T = 4.2 K, M: 14P-SM, Lit. [10.10,39]

CuBr

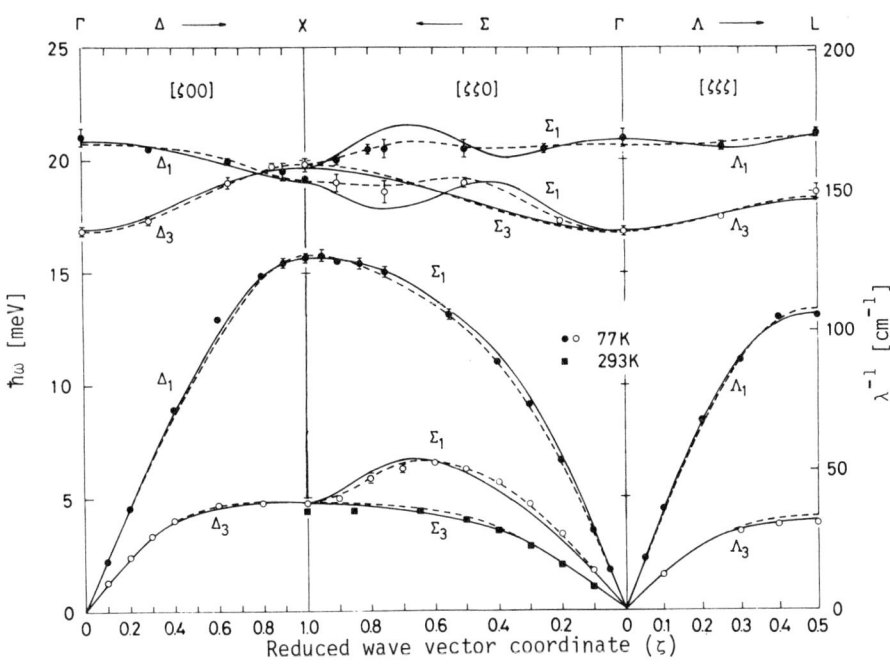

Fig. 10.14a. CuBr: $\omega(q)$ [Ref. 10.40, Fig. 3], T = 77 K, M: 14P-SM (Parameter set I: dashed lines, parameter set II: full lines), Lit.[10.39,41]

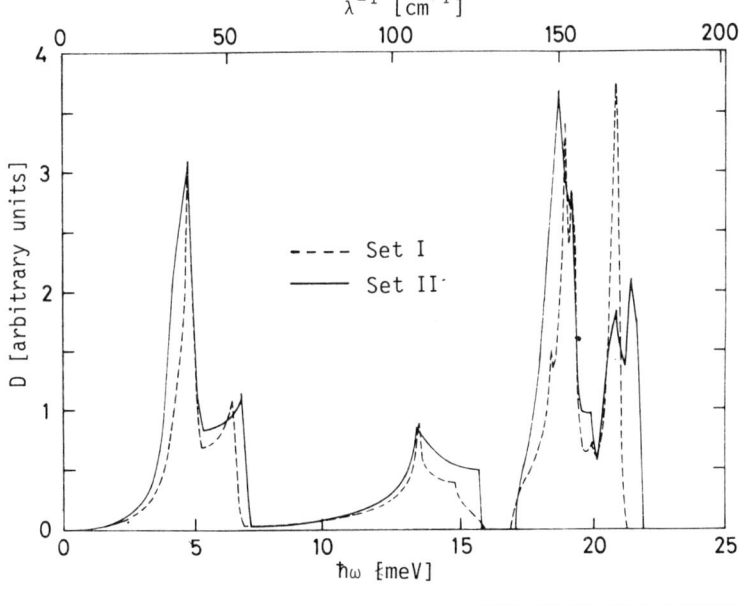

Fig. 10.14b. CuBr: [Ref. 10.40,Fig.4], T = 77 K, M: 14P-SM, Lit. [10.39, 41]

CuI

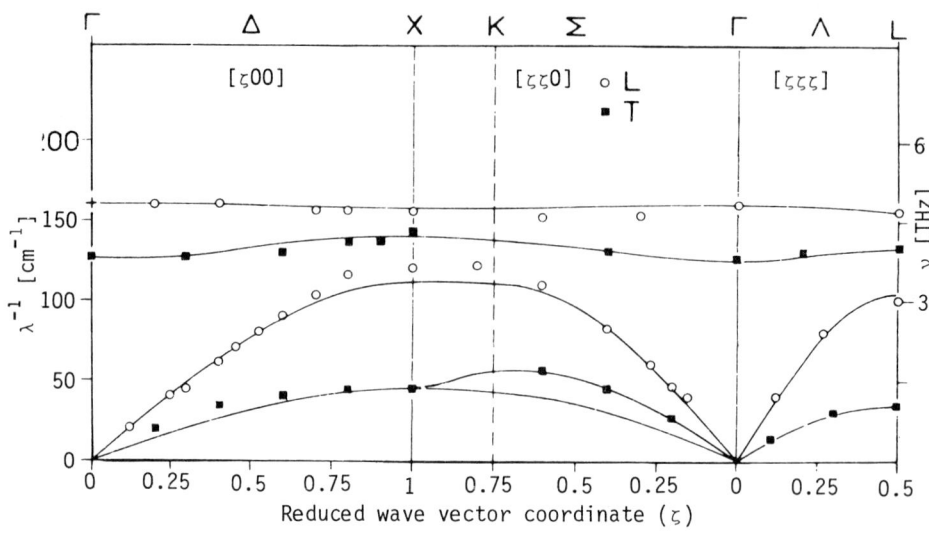

Fig. 10.15a. CuI: $\omega(\underline{q})$ [Ref. 10.42, Fig. 1], T = 293 K, M: 7P-RIM, Lit. [10.10,30,39]

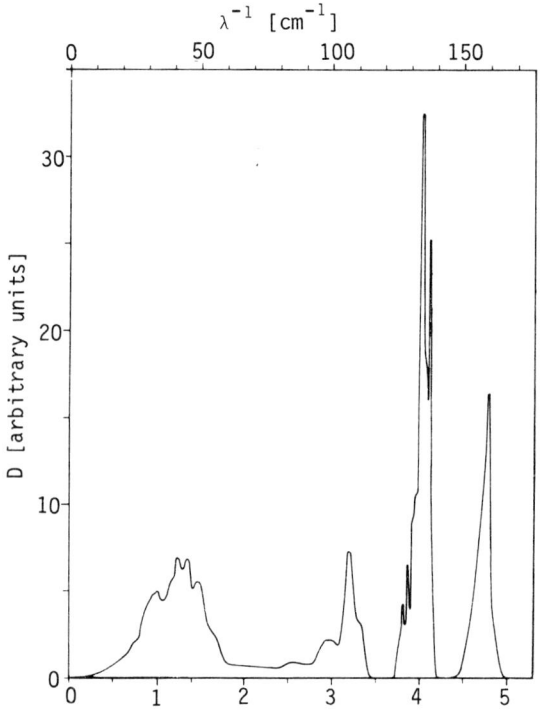

Fig. 10.15b. CuI: $D(\omega)$ [Ref. 10.42, Fig. 2], T = 293 K, M: 7P-RIM, Lit. [10.10,39]

11. Wurtzite Structure Crystals

The lattice dynamics of crystals with wurtzite structure belongs to the most difficult of diatomic crystals in general. As an example, the controversial situation in ZnO may be regarded. The delicate points are the 2NN forces which are different from those in crystals with zinc-blende structure (Chap. 10). Of particular interest is β-AgI for its relation to the fast ionic Ag conduction in α-AgI at high temperatures.

Crystal	Figures showing	
	Dispersion curve $\omega(\underline{q})$	Density of states $D(\omega)$
BeO	11.1a	11.1b
ZnO	11.2a 11.2b	
CdS	11.3a	11.3b
β-AgI	11.4a	11.4b

BeO

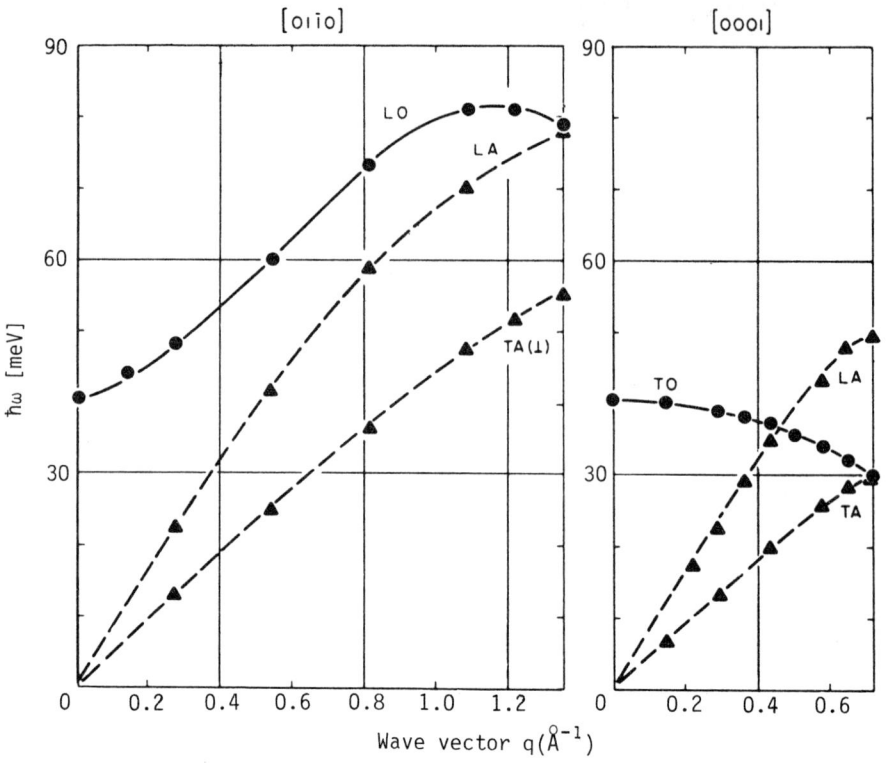

Fig. 11.1a. BeO: $\omega(\underline{q})$ [Ref. 11.1, Fig. 1], T = 300 K, Lit. [11.2-4]

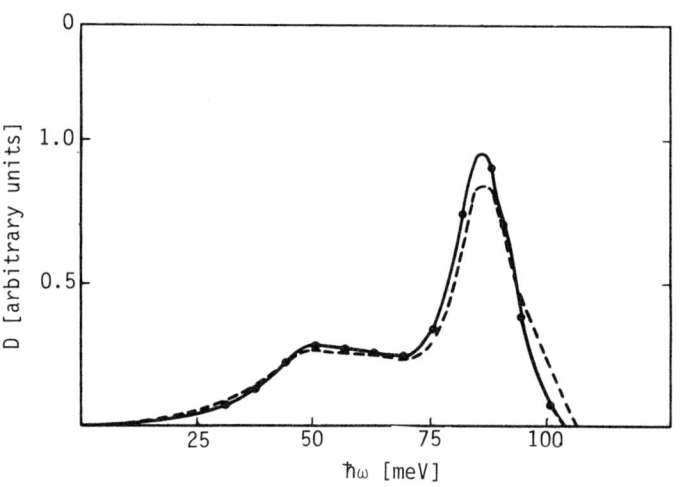

Fig. 11.1b. BeO: $D(\omega)$ [Ref. 11.4, Fig.7], T = 293 K

ZnO

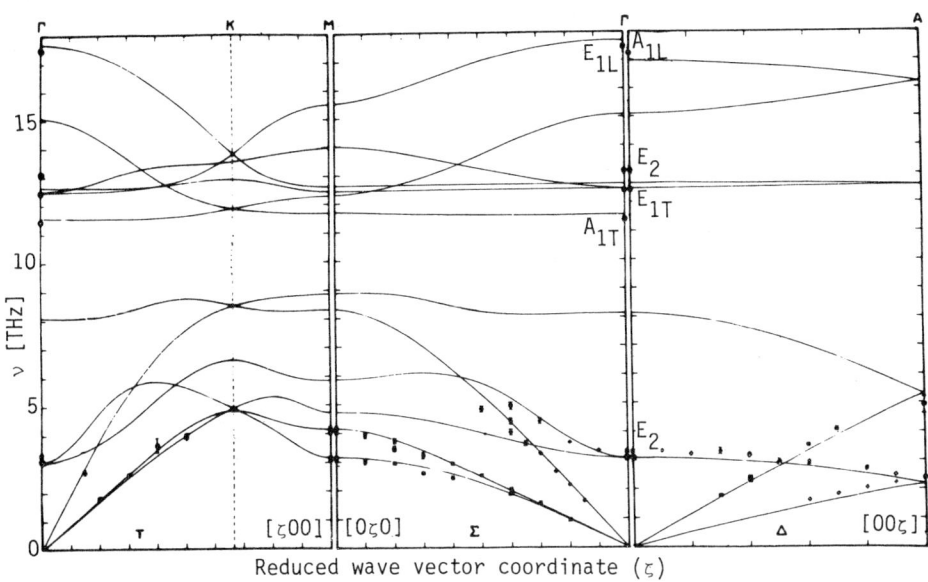

Fig. 11.2a. ZnO: $\omega(\underline{q})$ [Ref. 11.5, Fig. 1], T = 298 K, M: 7P-SM, Lit. [11.7,8]

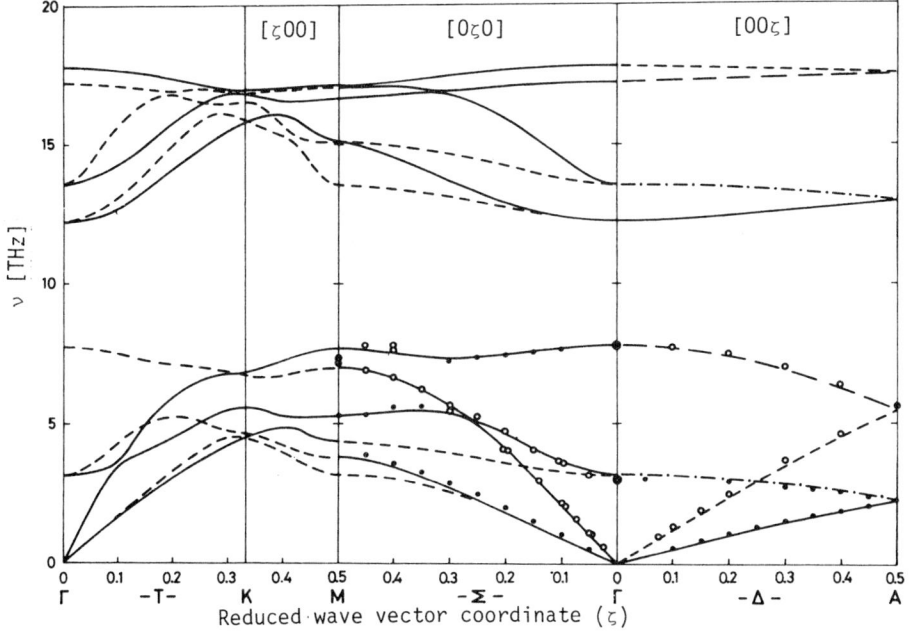

Fig. 11.2b. ZnO: $\omega(\underline{q})$ [Ref.11.6, Fig.1], T = 293 K, M: 10P-VSM, Lit.[11,7,8]

CdS

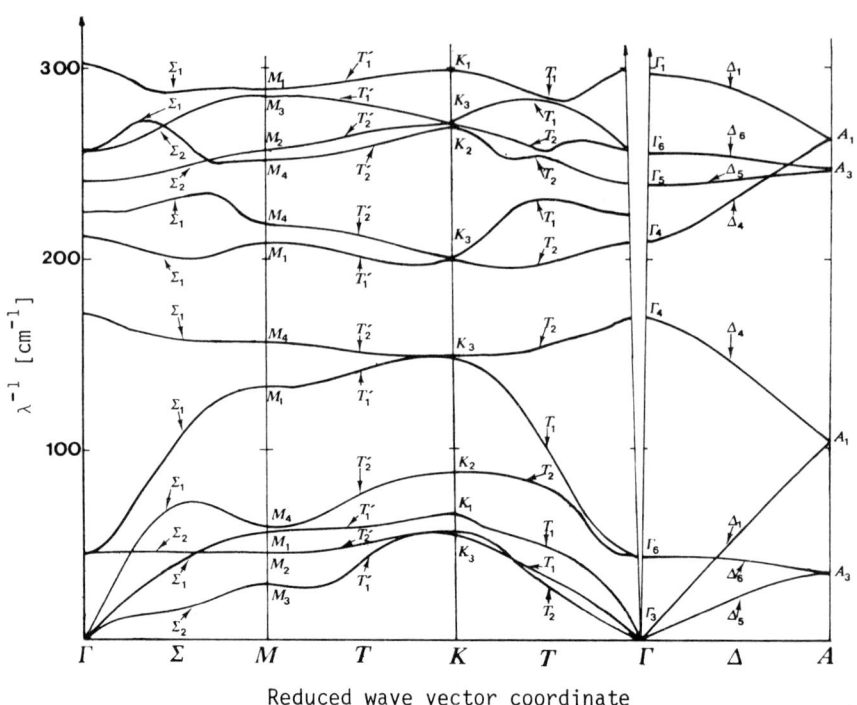

Fig. 11.3a. CdS: $\omega(\mathbf{q})$ [Ref. 11.9, Fig. 1], M: 10P-VDDM

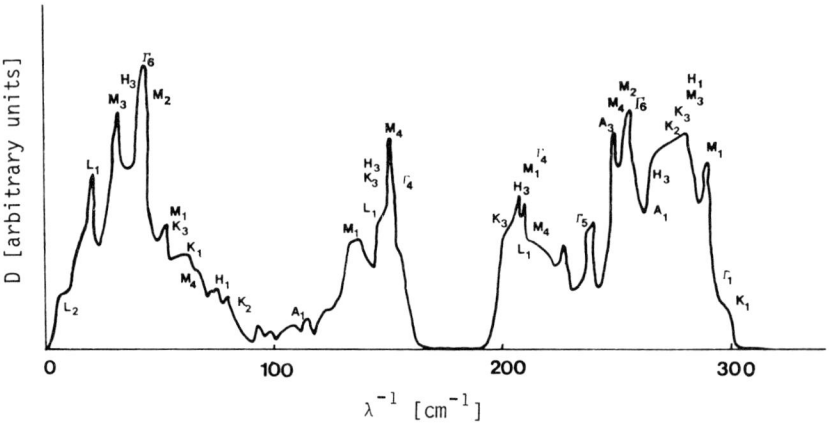

Fig. 11.3b. CdS: $D(\omega)$ [Ref. 11.9, Fig. 2], M: 10P-VDDM

β-AgI

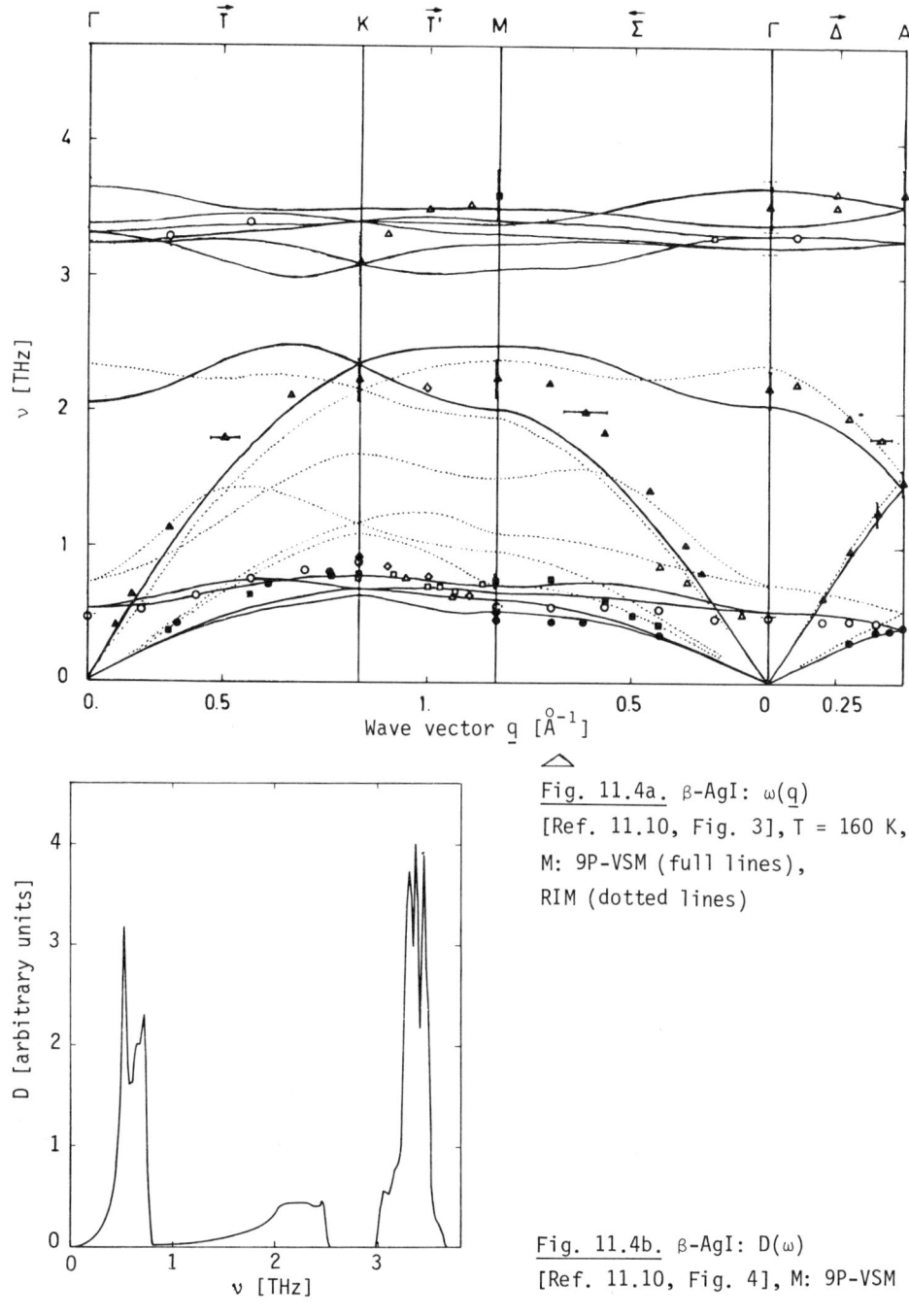

Fig. 11.4a. β-AgI: ω(q) [Ref. 11.10, Fig. 3], T = 160 K, M: 9P-VSM (full lines), RIM (dotted lines)

Fig. 11.4b. β-AgI: D(ω) [Ref. 11.10, Fig. 4], M: 9P-VSM

12. Fluorite Structure Crystals

CaF_2 and its structural homologues are, on the one side, normal ionic crystals, but on the other side have several members which show fast ionic conduction and, related to that, unusual anharmonic properties. The shell models used so far do not exhibit a clear relation to this latter feature in fluorite structured crystals.

The following systems are treated:

Crystal	Figures showing	
	Dispersion curve $\omega(\underline{q})$	Density of states $D(\omega)$
CaF_2	12.1a	12.1b
SrF_2	12.2a	12.2b
$SrCl_2$	12.3a	12.3b
BaF_2	12.4a	12.4b
PbF_2	12.5a	12.5b
Mg_2Sn	12.6	
Mg_2Pb	12.7a	12.7b
UO_2	12.8a	12.8b

Note added in proof: The dispersion curves of Na_2S (anti-fluorite structure) have been measured recently by W. Bührer and H. Bill and are reported in Helvetica Physica Acta *50*, 431 (1977).

CaF$_2$

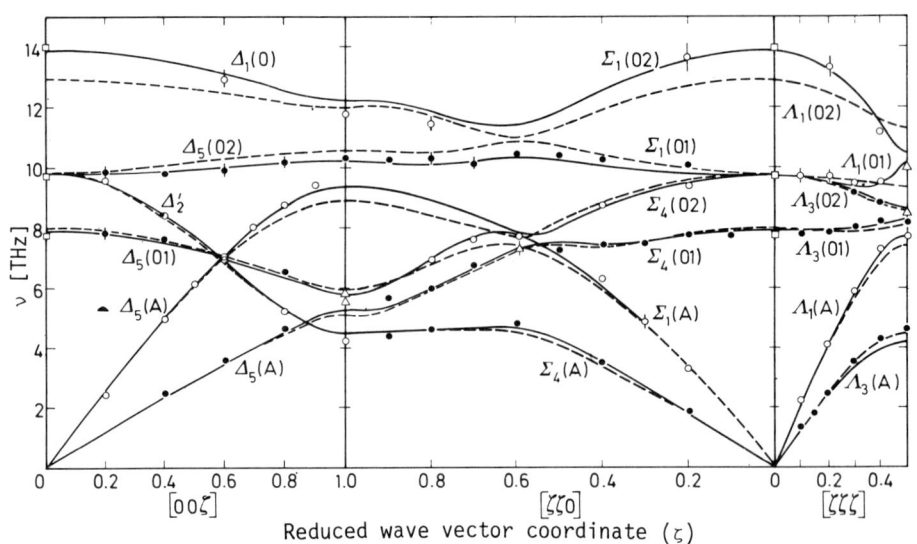

Fig. 12.1a. CaF$_2$: $\omega(\underline{q})$ [Ref. 12.1, Fig. 1], T = 295 K, M: 13P-SM (full lines), 7P-RIM (dashed lines)

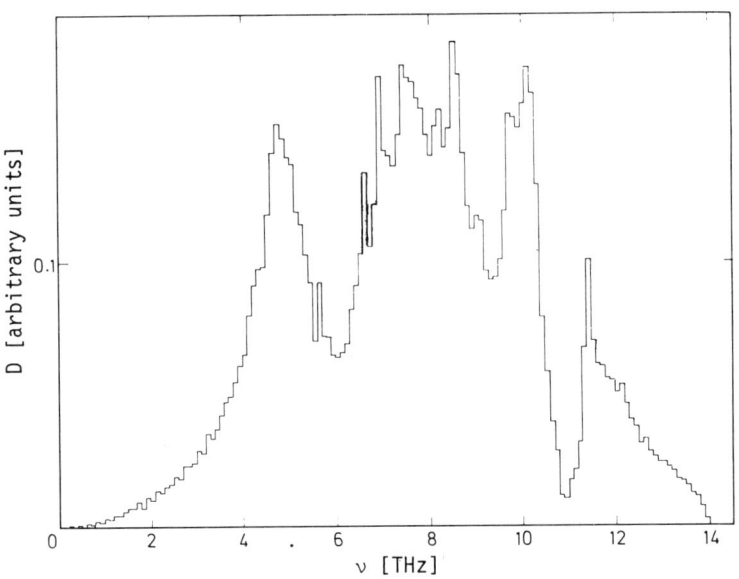

Fig. 12.1b. CaF$_2$: $D(\omega)$ [Ref. 12.1, Fig. 2], T = 295 K, M: 13P-SM

SrF$_2$

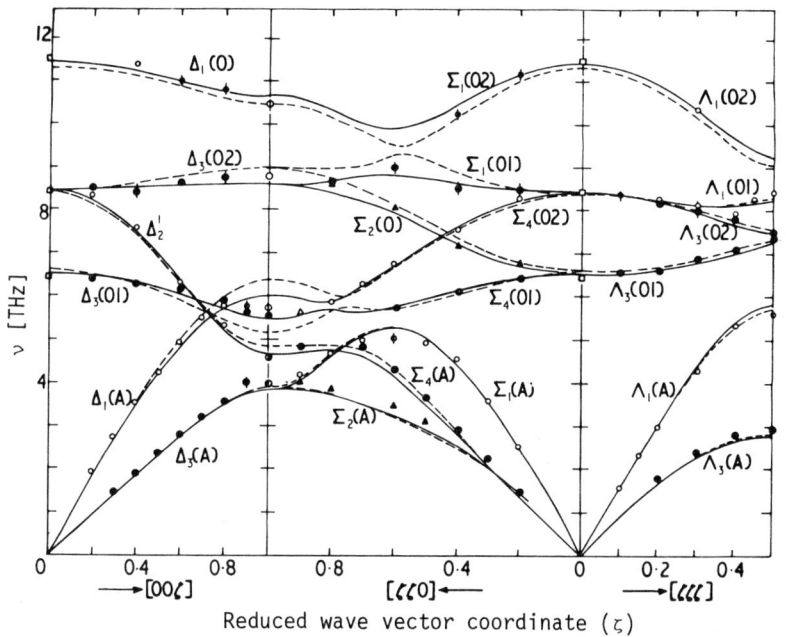

Fig. 12.2a. SrF$_2$: $\omega(\underline{q})$ [Ref. 12.2, Fig. 1], T = 295 K, M: 14P-SM (full lines), 10P-RIM (dashed lines)

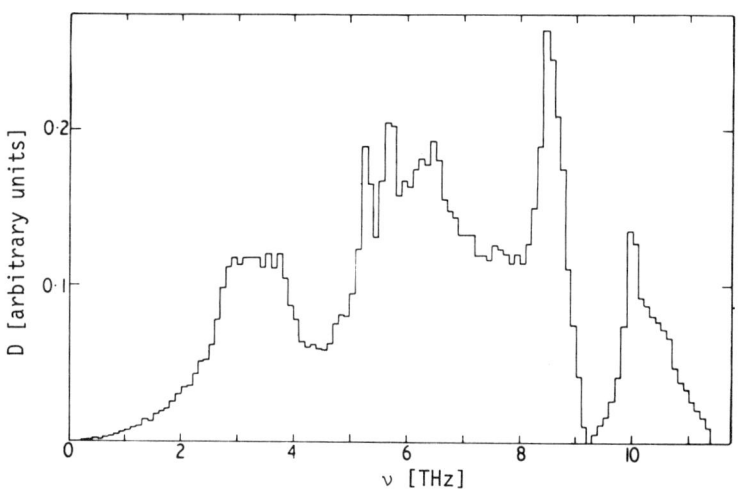

Fig. 12.2b. SrF$_2$: $D(\omega)$ [Ref. 12.2, Fig. 2], T = 295 K, M: 14P-SM

SrCl₂

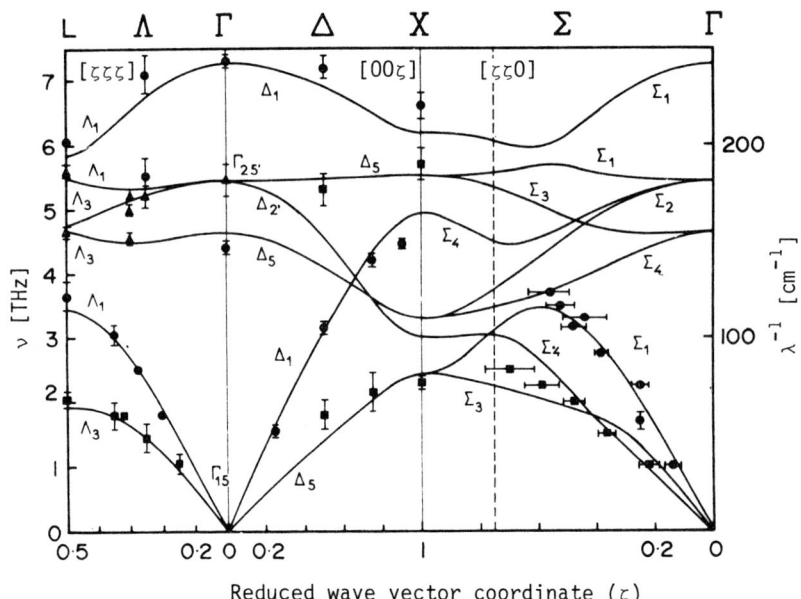

Fig. 12.3a. $SrCl_2$: $\omega(\underline{q})$ [Ref. 12.3, Fig. 1], T = RT, M: 8P-SM

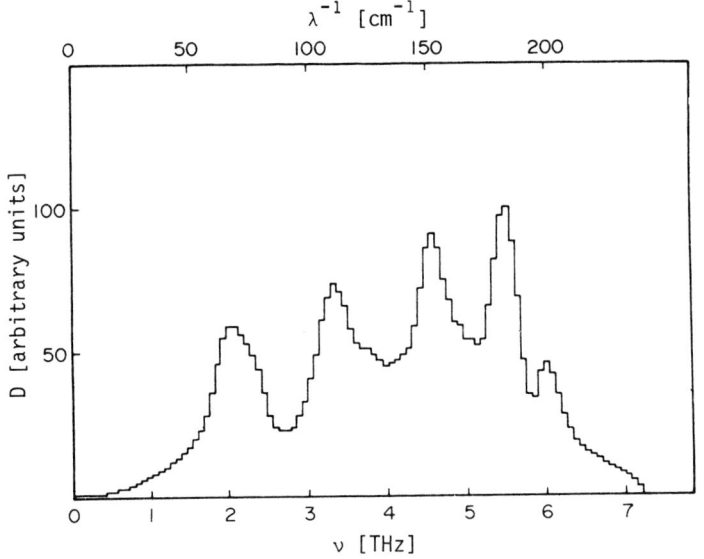

Fig. 12.3b. $SrCl_2$: $D(\omega)$ [Ref. 12.3, Fig. 2], T = RT, M: 8P-SM

BaF$_2$

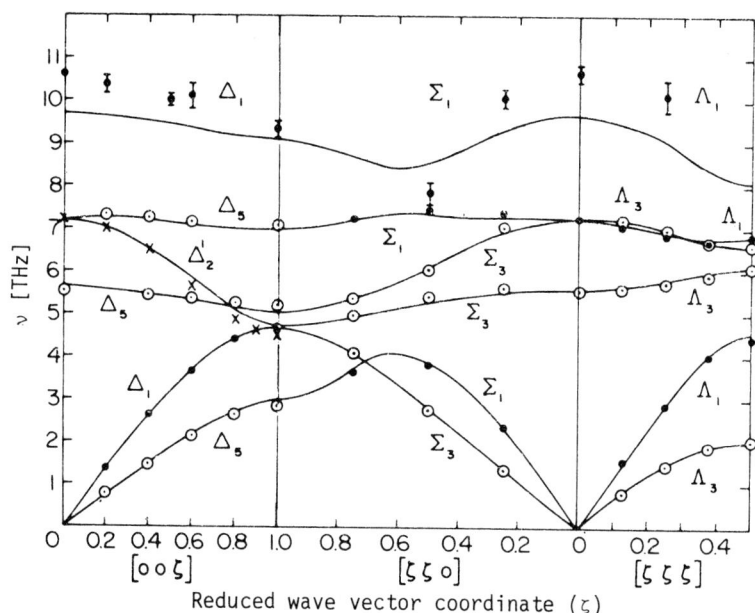

Fig. 12.4a. BaF$_2$: $\omega(\underline{q})$ [Ref. 12.4, Fig. 1], T = RT, M: 11P-SM

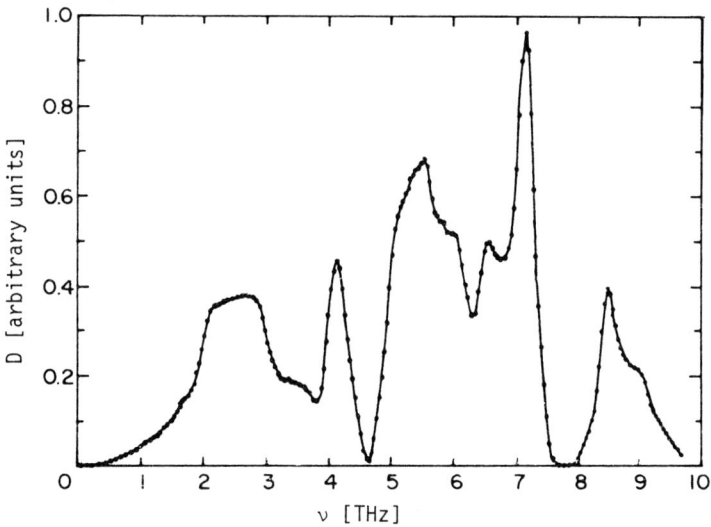

Fig. 12.4b. BaF$_2$: $D(\omega)$ [Ref. 12.4, Fig. 2], T = RT, M: 11P-SM

PbF$_2$

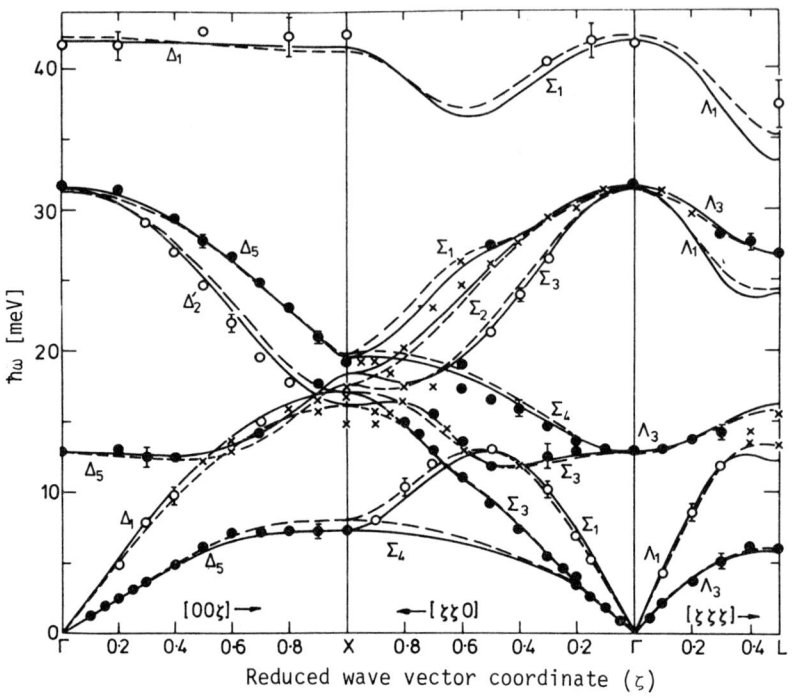

Fig. 12.5a. PbF$_2$: $\omega(\underline{q})$ [Ref. 12.5, Fig. 1], T = 10 K, M: 13P-SM (full lines), 8P-SM (dashed lines)

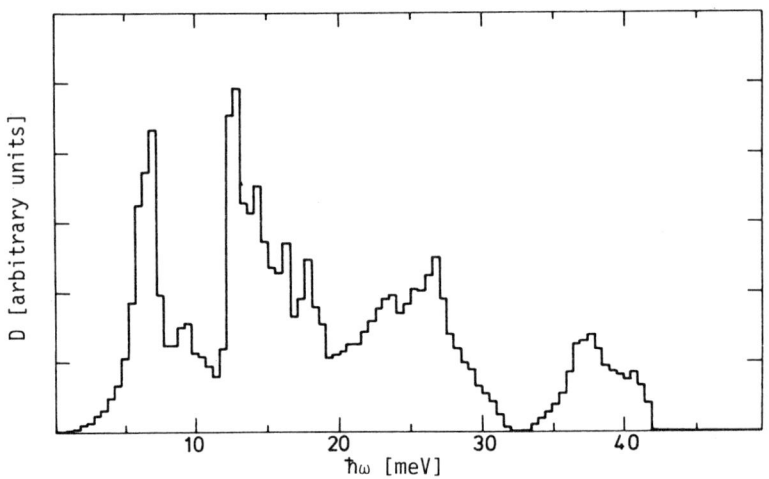

Fig. 12.5b. PbF$_2$: D(ω) [Ref. 12.5, Fig. 2], T = 10 K, M: 13P-SM

Mg$_2$Sn

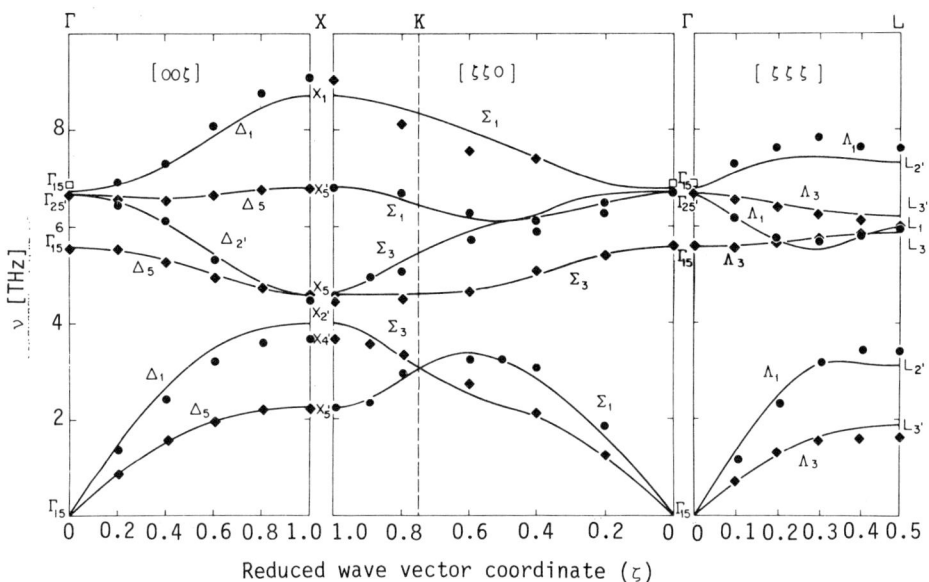

Fig. 12.6. Mg$_2$Sn: $\omega(\underline{q})$ [Ref. 12.6, Fig. 3], T = RT, M: 10P-SM

Mg$_2$Pb

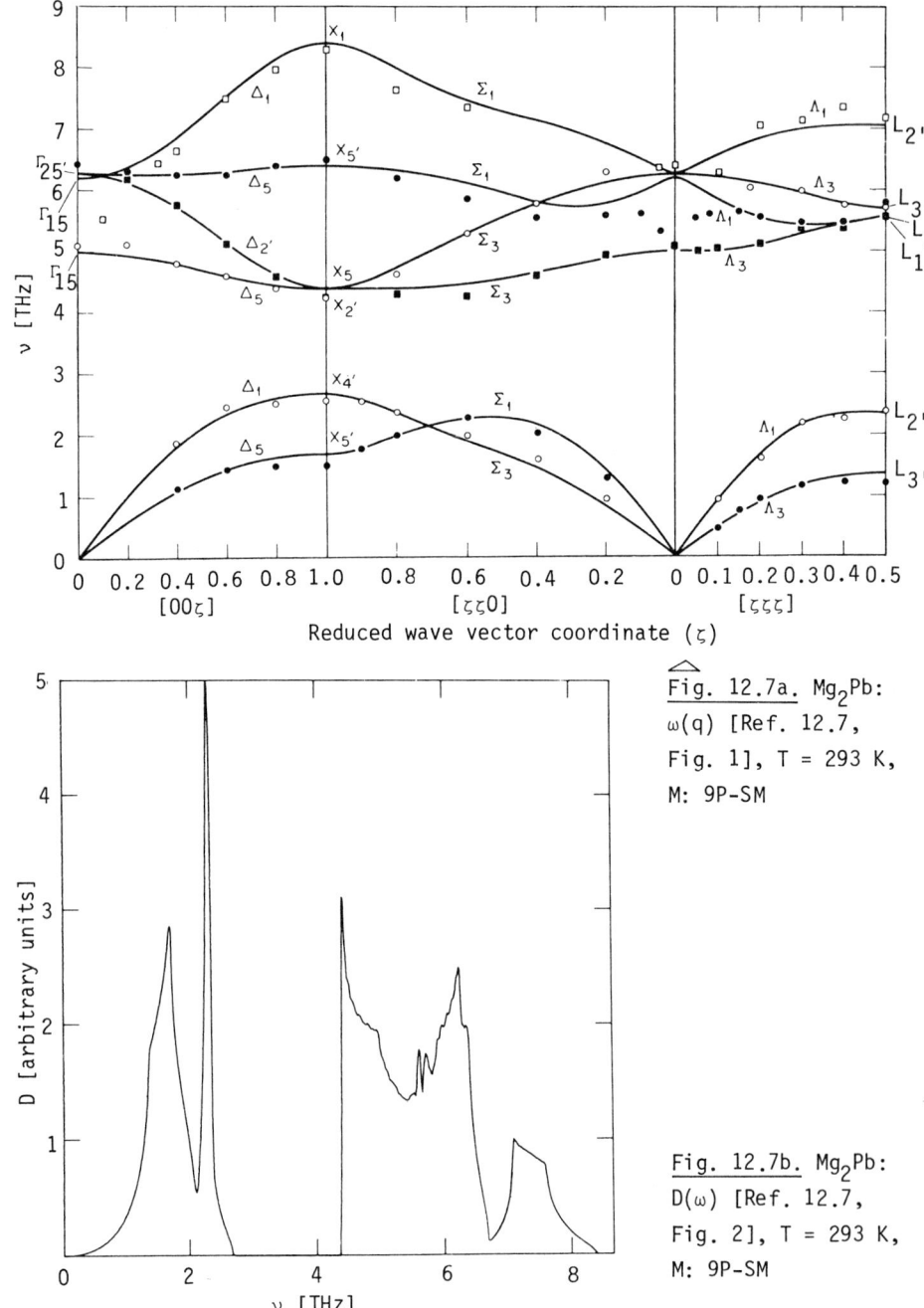

Fig. 12.7a. Mg$_2$Pb: $\omega(q)$ [Ref. 12.7, Fig. 1], T = 293 K, M: 9P-SM

Fig. 12.7b. Mg$_2$Pb: $D(\omega)$ [Ref. 12.7, Fig. 2], T = 293 K, M: 9P-SM

UO$_2$

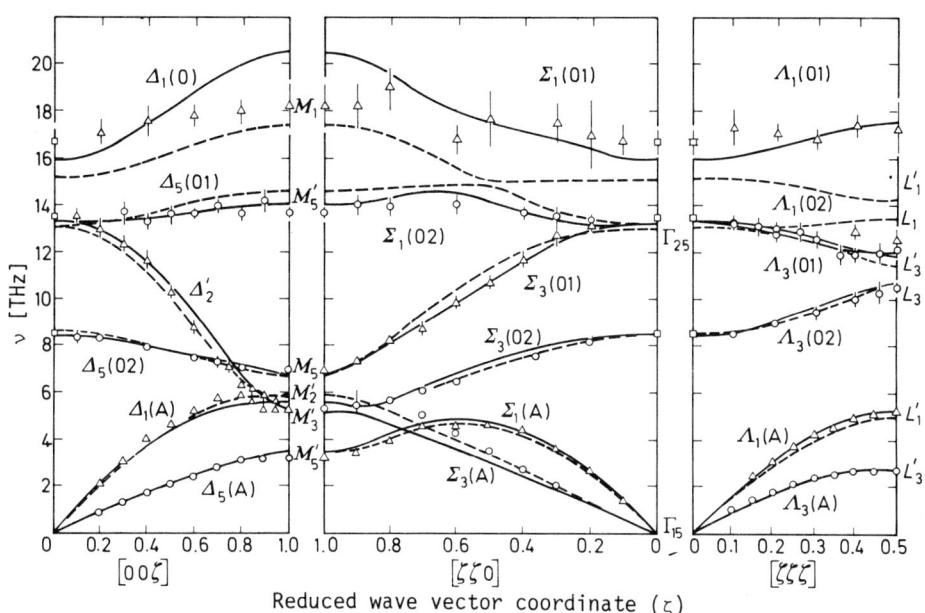

Fig. 12.8a. UO$_2$: $\omega(\underline{q})$ [Ref. 12.8, Fig. 4], T = 296 K, M: 11P-SM, (solid lines), 7P-RIM (dashed lines), Lit. [12.9]

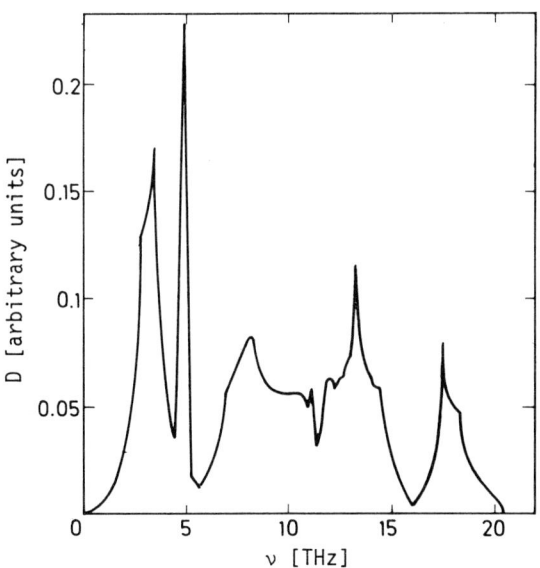

Fig. 12.8b. UO$_2$: D(ω) [Ref. 12.8, Fig. 5], M: 11P-SM, (solid lines), 7P-RIM (dashed lines)

13. Rutile Structure Crystals

The crystals in this group show many interesting features which one might expect in a relatively open structure, in particular phase transitions as a function of pressure and temperature. The description of the dispersion curves by shell models with many adjustable parameters is very unsatisfactory, at least in the oxides, while the fluorides exhibit a simpler behavior.

The following systems are treated:

	Figures showing	
Crystal	Dispersion curve $\omega(\underline{q})$	Density of states $D(\omega)$
TiO_2	13.1a,b	13.1c
MgF_2	13.2a	13.2b
MnF_2	13.3a	13.3b
FeF_2	13.4	
CoF_2	13.5	
SnO_2	13.6	

TiO$_2$

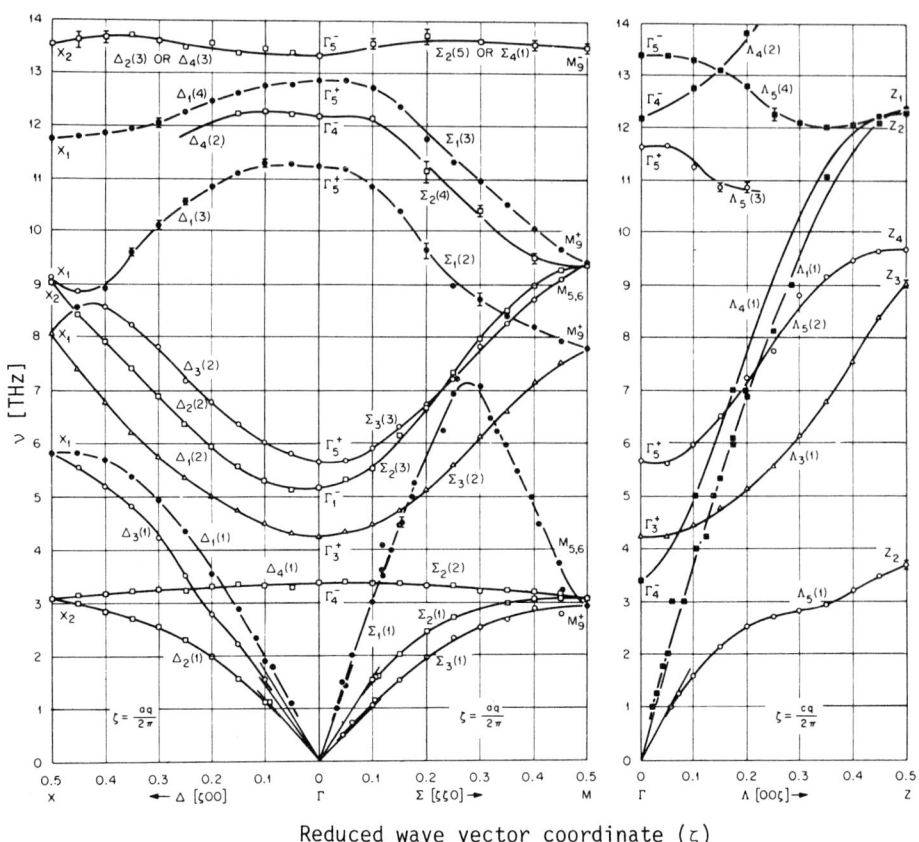

Reduced wave vector coordinate (ζ)

<u>Fig. 13.1a.</u> TiO$_2$: $\omega(\underline{q})$ [Ref. 13.1, Fig. 5], T = RT, M: -, Lit. [13.2]

TiO₂

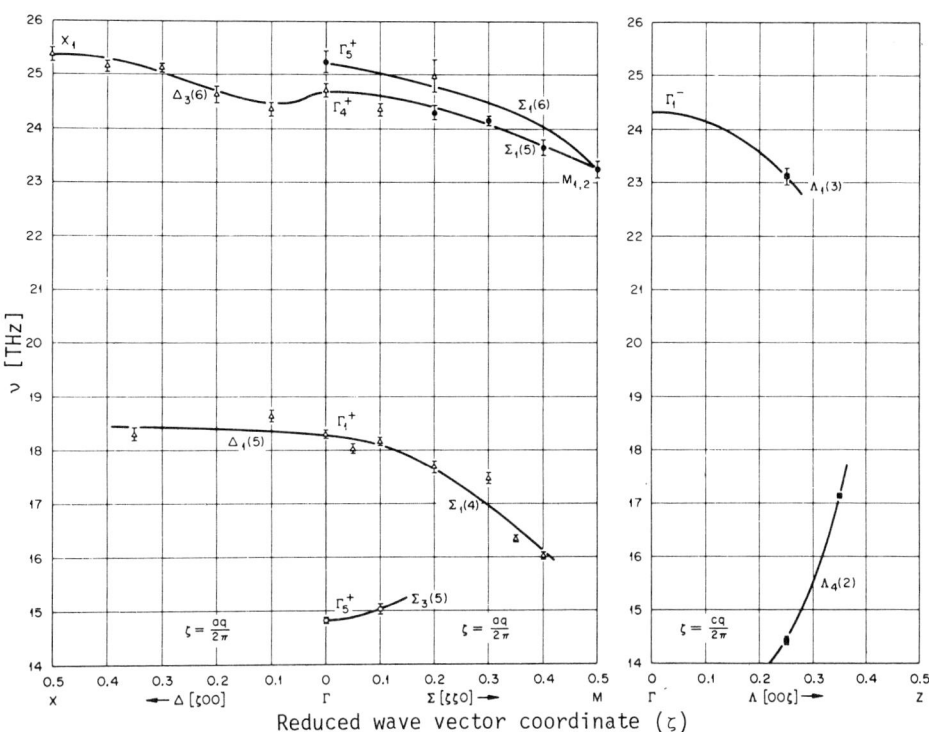

Fig. 13.1b. TiO$_2$: $\omega(\underline{q})$ [Ref. 13.1, Fig. 6], T = RT, M: -

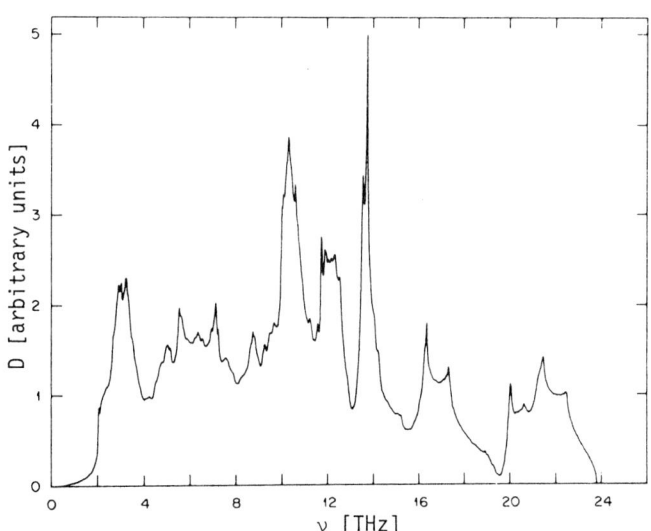

Fig. 13.1c. TiO$_2$: $D(\omega)$ [Ref. 13.1, Fig. 10], M: 18P-SM

MgF$_2$

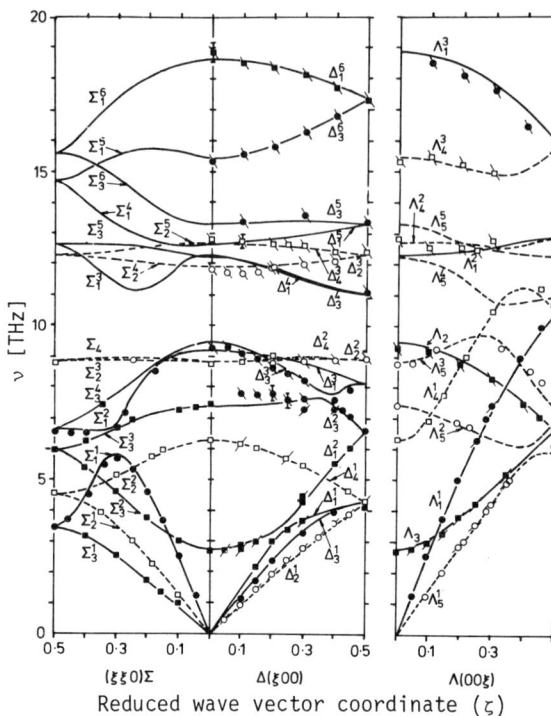

Fig. 13.2a. MgF$_2$: $\omega(\underline{q})$ [Ref. 13.9, Fig. 2], T = 296K, M: 14P-SM, Lit. [13.3,5]

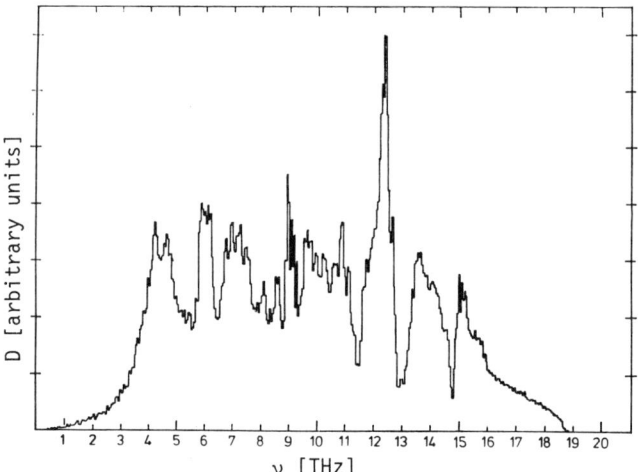

Fig. 13.2b. MgF$_2$: $D(\omega)$ [Ref. 13.9, Fig. 3], M: 14P-SM

MnF$_2$

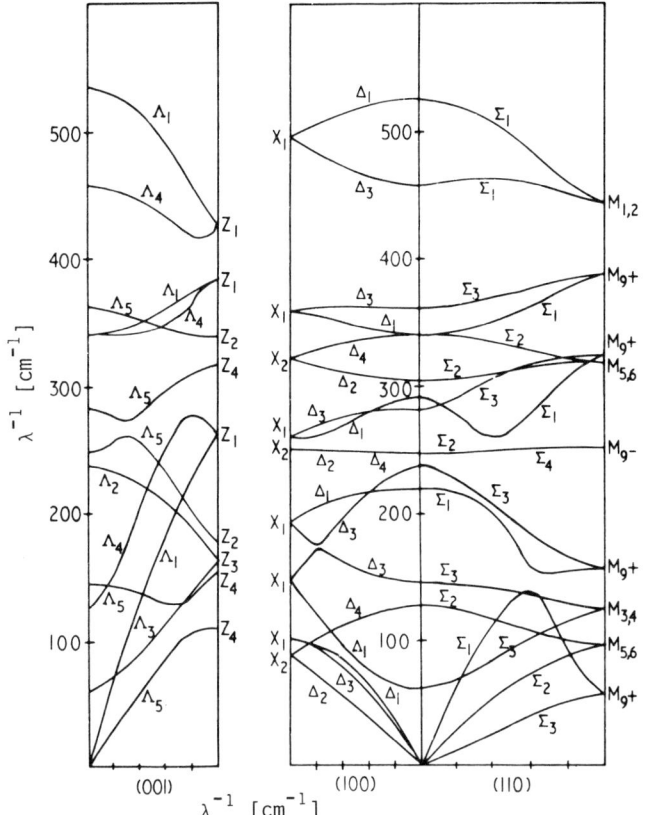

Fig. 13.3a. MnF$_2$: $\omega_c(\underline{q})$ [Ref. 13.4, Fig. 2], M: 11P-SM

Note added in proof: The dispersion curves measured by C.A. Rotter, J.G. Traylor, and H.G. Smith are presented in a review article by H.G. Smith and N. Wakabayashi published in *Dynamics of Solids and Liquids by Neutron Scattering* edited by T. Springer and S. Lovesey; Topics in Current Physics, Vol. 3 (Springer, Berlin, Heidelberg, New York 1977) Fig. 2.8.

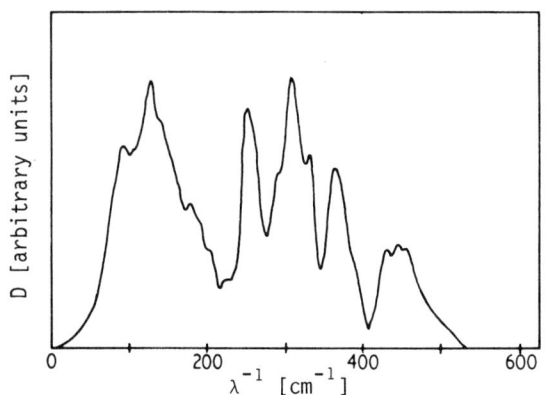

Fig. 13.3b. MnF$_2$: $D(\omega)$ [Ref. 13.4, Fig. 3a], M: 11P-SM

FeF$_2$

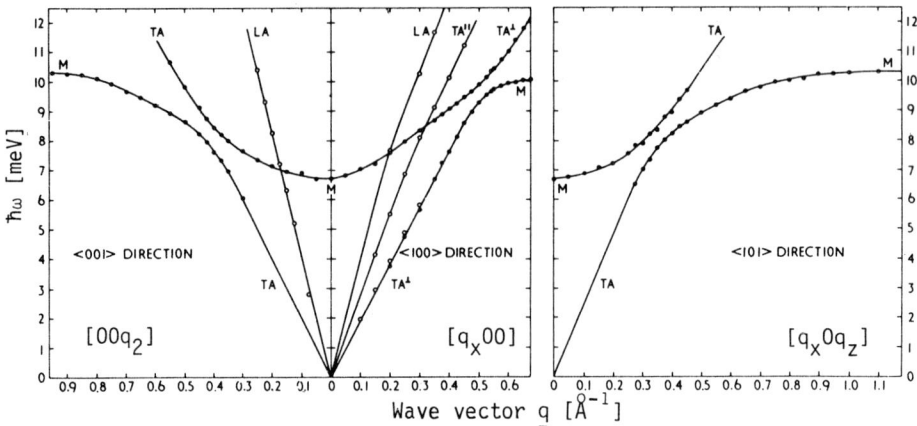

Fig. 13.4. FeF$_2$: $\omega(q)$ [Ref. 13.6, Fig. 1], T = 4.2K, M: -

CoF$_2$

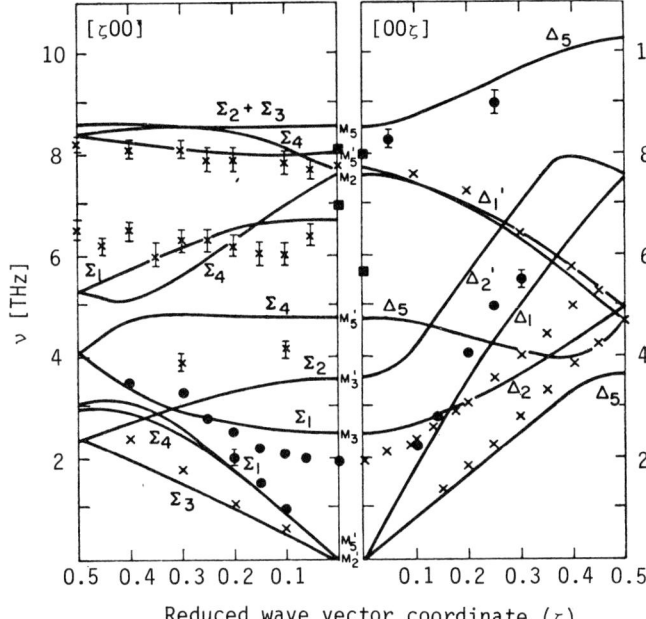

Fig. 13.5. CoF$_2$: $\omega(q)$ [Ref. 13.7, Fig. 5], T = 80K, M: 3P-RIM

SnO$_2$

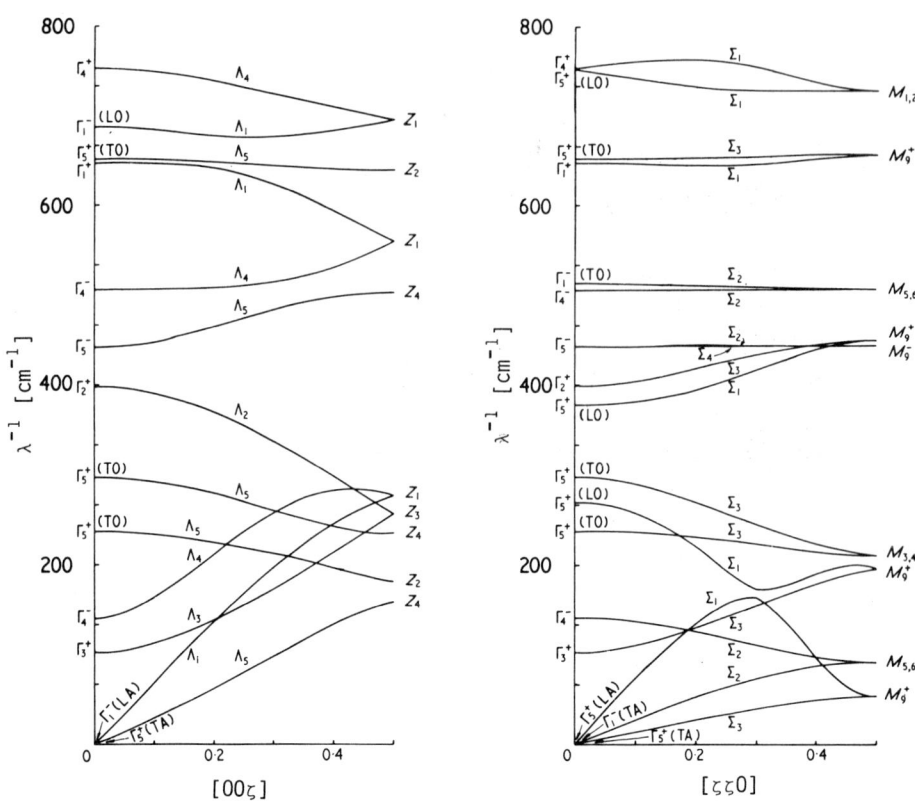

Reduced wave vector coordinate (ζ)

Fig. 13.6. SnO$_2$: $\omega_c(\underline{q})$ [Ref. 13.8, Fig. 4], M: 9P-RIM

14. ABO₃ and ABX₃ Crystals

In this chapter the phonon data of all measured perovskitic crystals are collected together with other crystals which belong to the chemical brutto formula ABX_3 (X = O,H,F). This, unavoidably, includes so-called molecular crystals like $NaNO_3$ but may be justified from a technical point of view. Since these crystals show many different types of phase transitions together with a complex dielectric and anharmonic behavior it is not surprising that the lattice dynamics of them is very complicated and offers many unsolved problems. An exception are the cubic perovskites where the pioneering work of COWLEY [14.12] and STIRLING [14.11] led to a basic understanding of interatomic forces in these substances. A recent microscopic model [14.7,8] emphasized the importance of the oxygen polarizability for the ferroelectric phase transition in ABO_3 perovskites.

The following systems are treated:

Crystal	Figures showing		Crystal	Figures showing	
	Dispersion curve $\omega(\underline{q})$	Density of states $D(\omega)$		Dispersion curve $\omega(\underline{q})$	Density of states $D(\omega)$
$LiNbO_3$	14.1		$BaLiH_3$		14.8
$KNbO_3$	14.2		$LaAlO_3$	14.9	
$KTaO_3$	14.3a-d		$PbTiO_3$	14.10a,b	
$KMnF_3$	14.4a,b		$CsNiF_3$	14.11	
$SrTiO_3$	14.5a	14.5b	$KCoF_3$	14.12	
$SrLiH_3$		14.6	$CaCO_3$	14.13	
$BaTiO_3$ (cubic)	14.7a,b		$NaNO_3$	14.14	
$BaTiO_3$ (tetrag.)	14.7c-e		KNO_3	14.15	

LiNbO₃

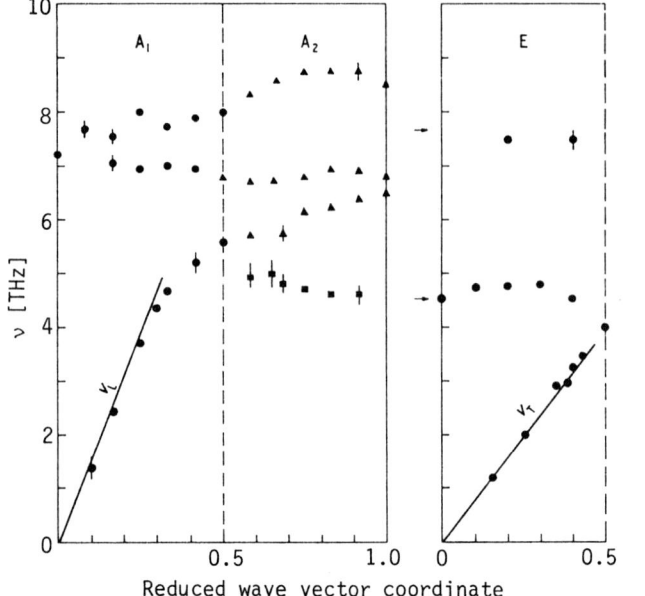

Fig. 14.1. LiNbO$_3$: $\omega(\underline{q})$ [Ref. 14.1, Fig. 1], T = RT, $\underline{q}$ = [00ζ]

KNbO$_3$

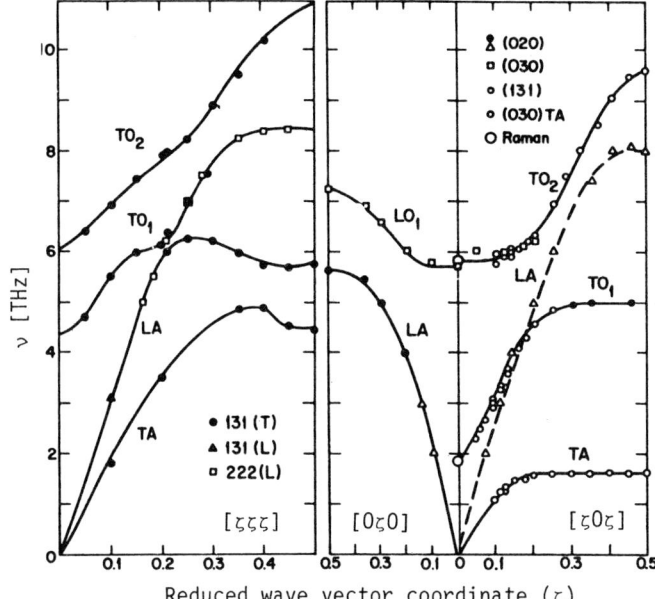

Fig. 14.2. KNbO$_3$: ω(q) [Ref. 14.2, Fig. 4], T = 300 K, M: -, Lit. [14.3]

KTaO₃

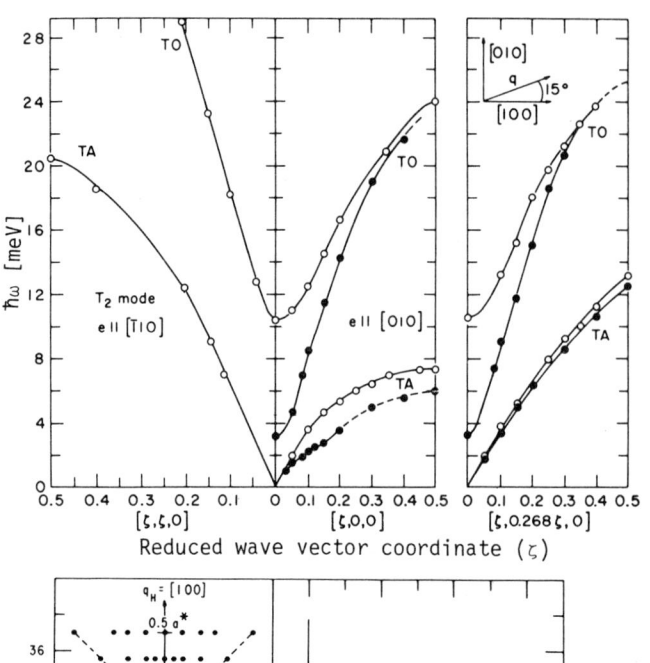

Fig. 14.3a. KTaO$_3$: ω(q) [Ref. 14.4, Fig. 2], T = 300K (open circles), T = 20K (solid circles), M: -, Lit. [14.5-8]

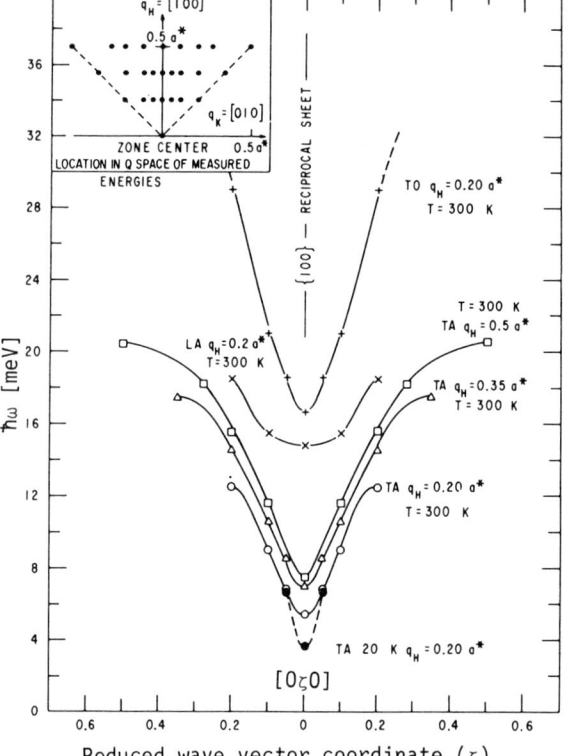

Fig. 14.3b. KTaO$_3$: ω(q) [Ref. 14.4, Fig. 3], M: -, Lit. [14.5-8]

KTaO$_3$

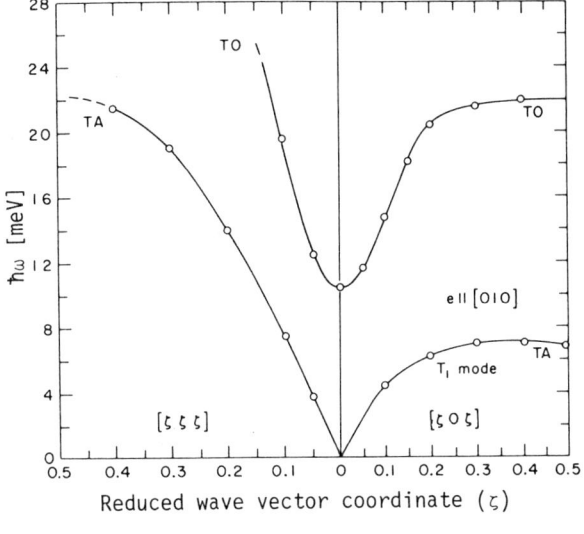

Fig. 14.3c. KTaO$_3$: $\omega(\underline{q})$ [Ref. 14.4, Fig. 4], T = 300 K, M: -, Lit. [14.5-8]

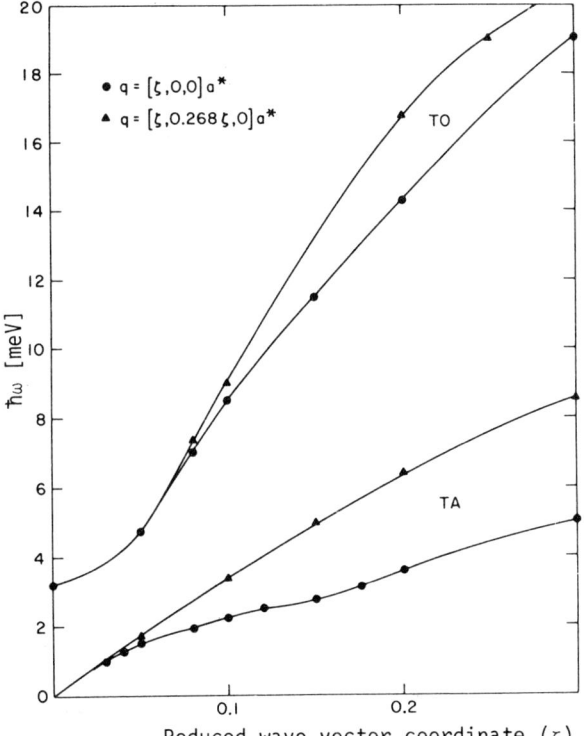

Fig. 14.3d. KTaO$_3$: $\omega(\underline{q})$ [Ref. 14.4, Fig. 5], T = 20 K, M: -, Lit. [14.5-8]

KMnF$_3$

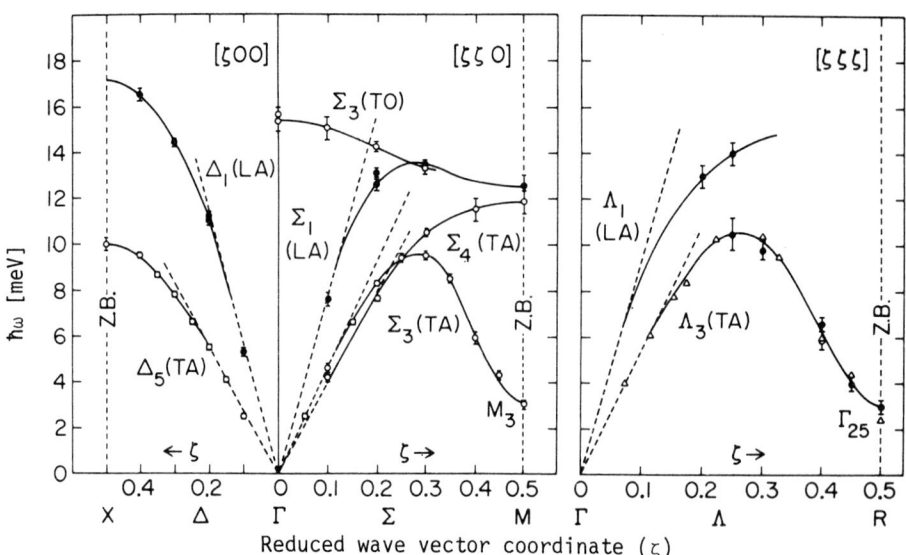

Fig. 14.4a. KMnF$_3$: $\omega(\underline{q})$ [Ref.14.9, Fig.2], T = 295K, M: -, Lit. [14.6,10]

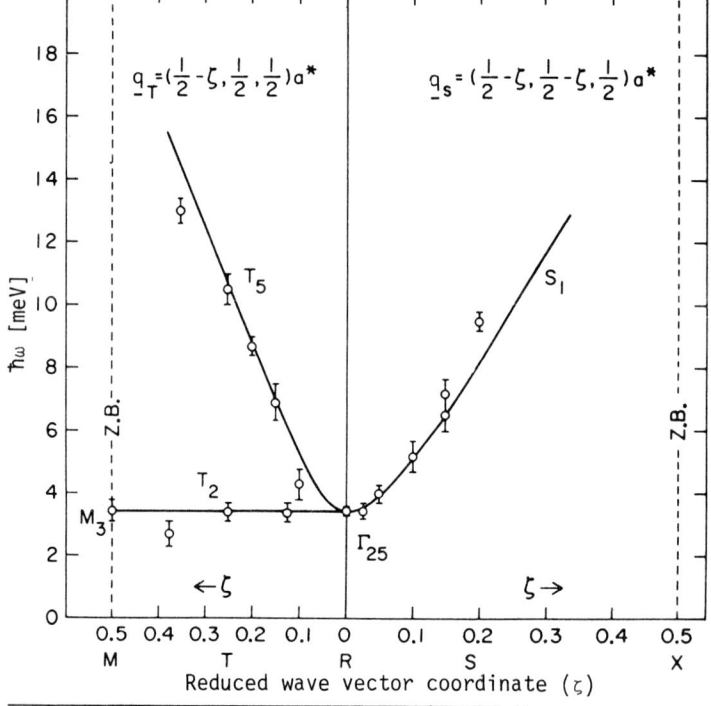

Fig. 14.4b. KMnF$_3$: $\omega(\underline{q})$ [Ref. 14.9, Fig. 3], T = 295 K, M: -, Lit. [14.6,10]

SrTiO$_3$

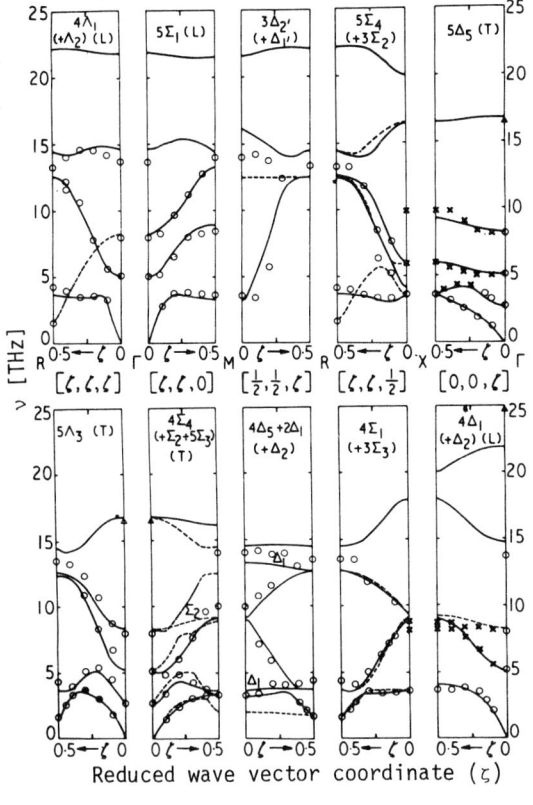

Fig. 14.5a. SrTiO$_3$: $\omega(\underline{q})$ [Ref. 14.11, Fig. 5], T = 297K, M: 14P-SM, crosses [14.12], Lit. [14.6,13-18]

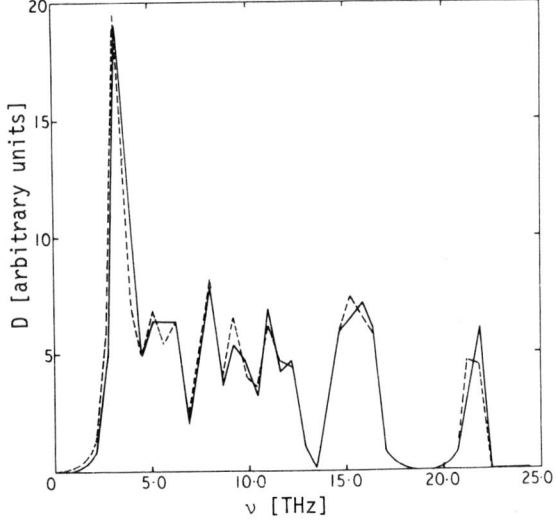

Fig. 14.5b. SrTiO$_3$: $D(\omega)$ [Ref. 14.11, Fig. 6], T = 297K (full lines), 90K (dashed lines), M: 14P-SM

SrLiH$_3$

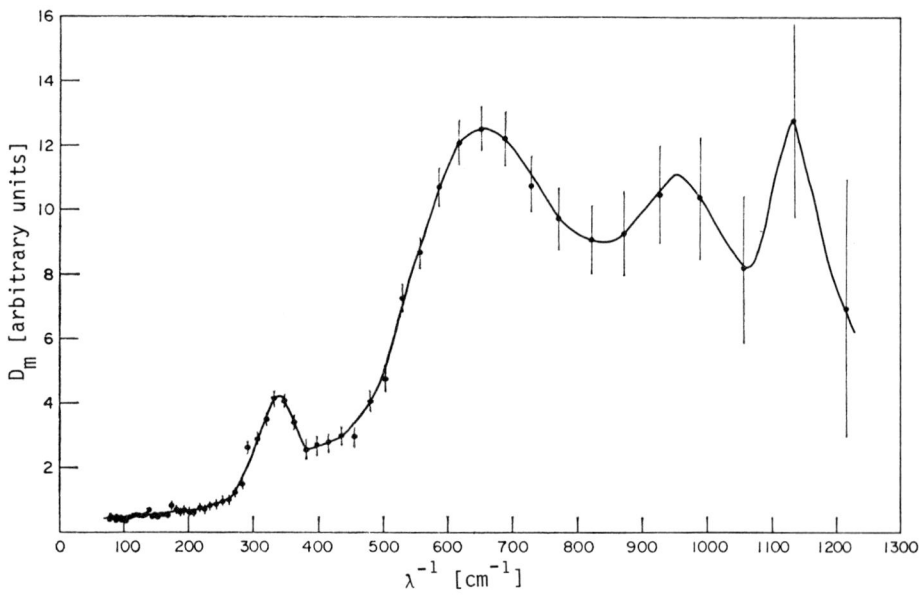

Fig. 14.6. SrLiH$_3$: D(ω) [Ref. 14.19, Fig. 4]

BaTiO₃ (cubic)

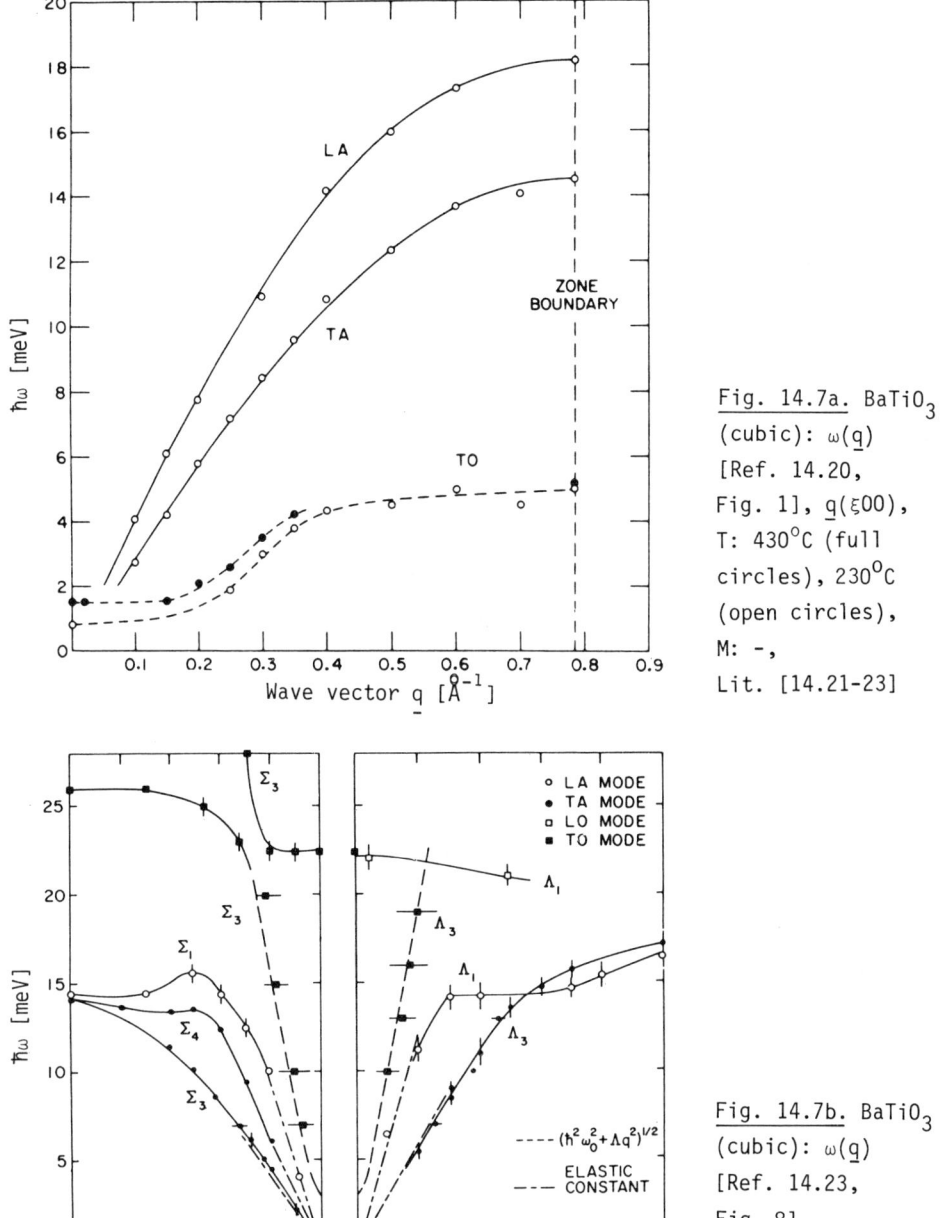

Fig. 14.7a. BaTiO$_3$ (cubic): $\omega(\underline{q})$ [Ref. 14.20, Fig. 1], $\underline{q}(\xi 00)$, T: 430°C (full circles), 230°C (open circles), M: -, Lit. [14.21-23]

Fig. 14.7b. BaTiO$_3$ (cubic): $\omega(\underline{q})$ [Ref. 14.23, Fig. 8], T = 150°C, M: - Lit. [14.20-22]

BaTiO$_3$ (tetragonal)

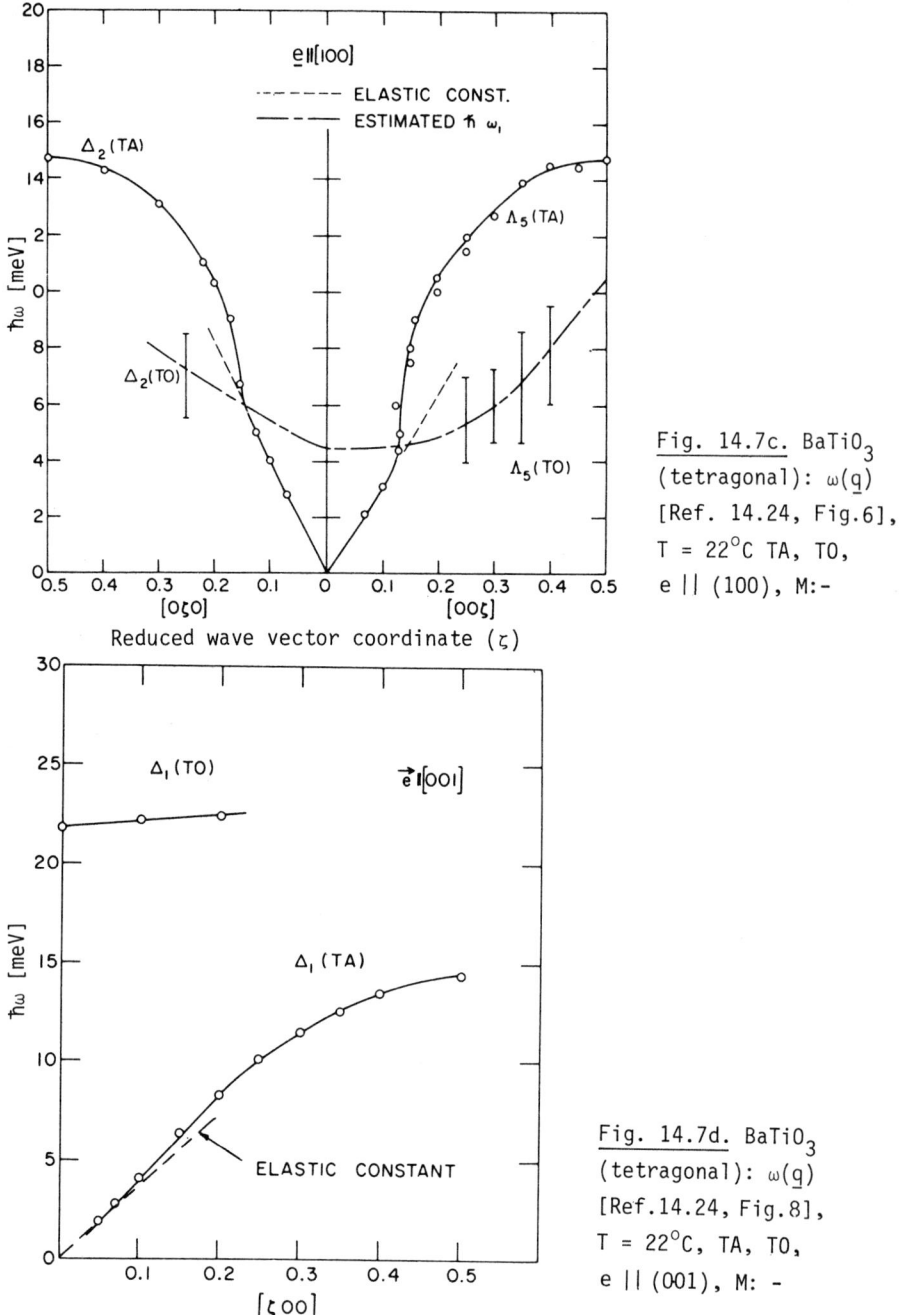

Fig. 14.7c. BaTiO$_3$ (tetragonal): ω(q) [Ref. 14.24, Fig.6], T = 22°C TA, TO, e || (100), M:-

Fig. 14.7d. BaTiO$_3$ (tetragonal): ω(q) [Ref.14.24, Fig.8], T = 22°C, TA, TO, e || (001), M: -

BaTiO$_3$

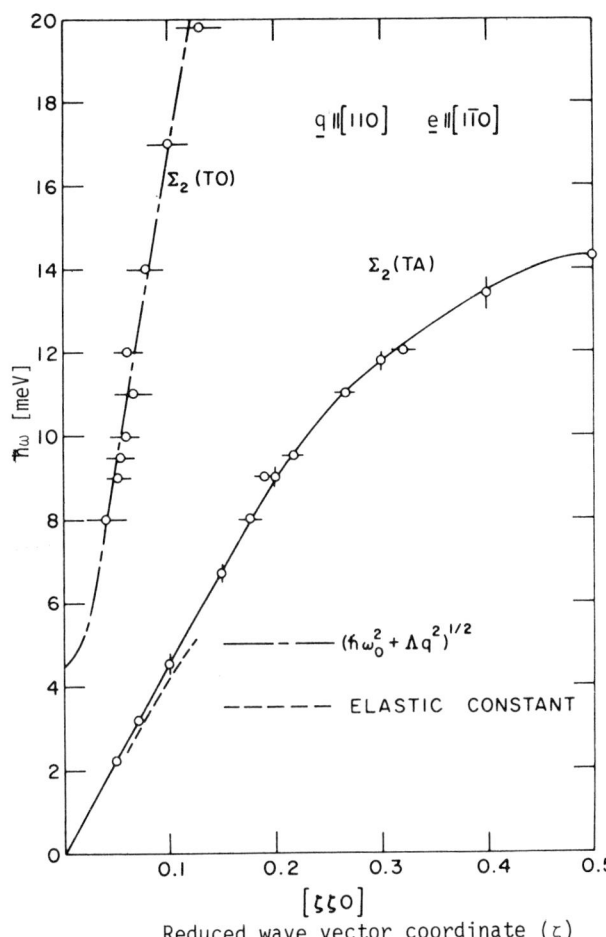

Fig. 14.7e. BaTiO$_3$ (tetragonal) $\omega(q)$ [Ref.14.24,Fig.3], T = 22°C, TA, TO, e || (110), M: -

BaLiH$_3$

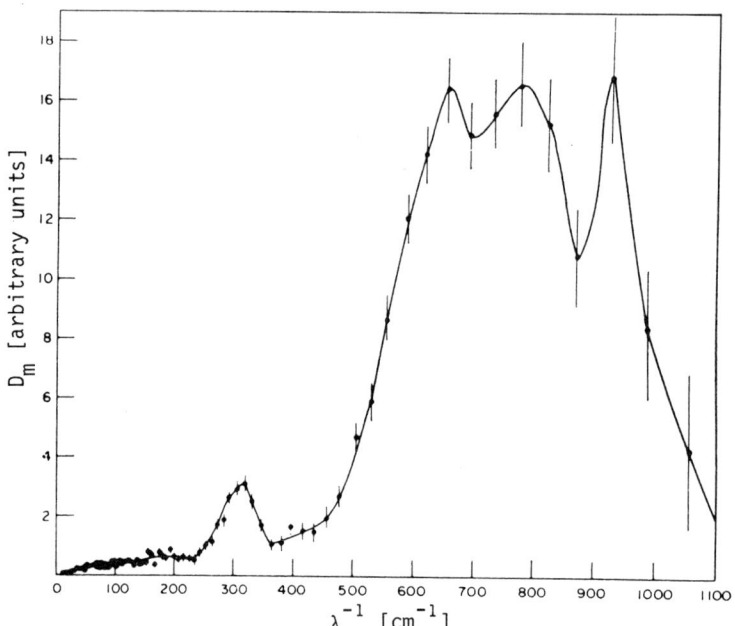

Fig. 14.8. BaLiH$_3$: D(ω) [Ref. 14.19, Fig. 3]

LaAlO$_3$

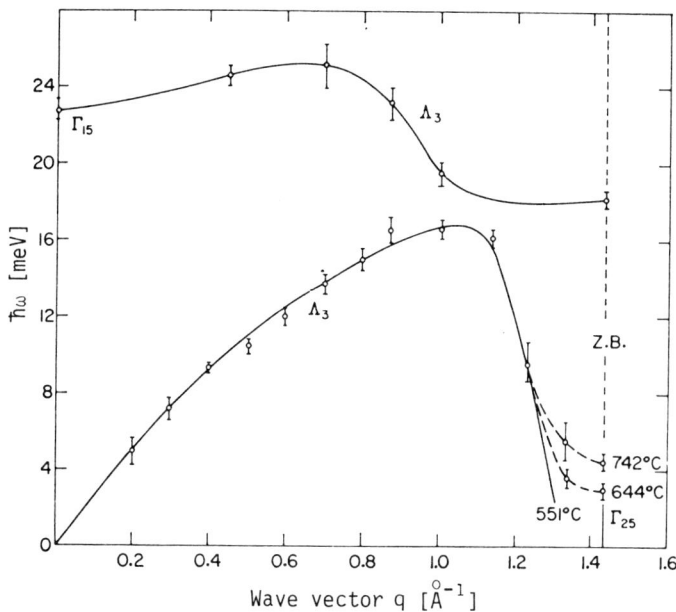

Fig. 14.9. LaAlO$_3$: $\omega(\underline{q})$ [Ref. 14.25, Fig. 3], T: 551°C, $\underline{q}$ ∥ (111), M: −

PbTiO$_3$

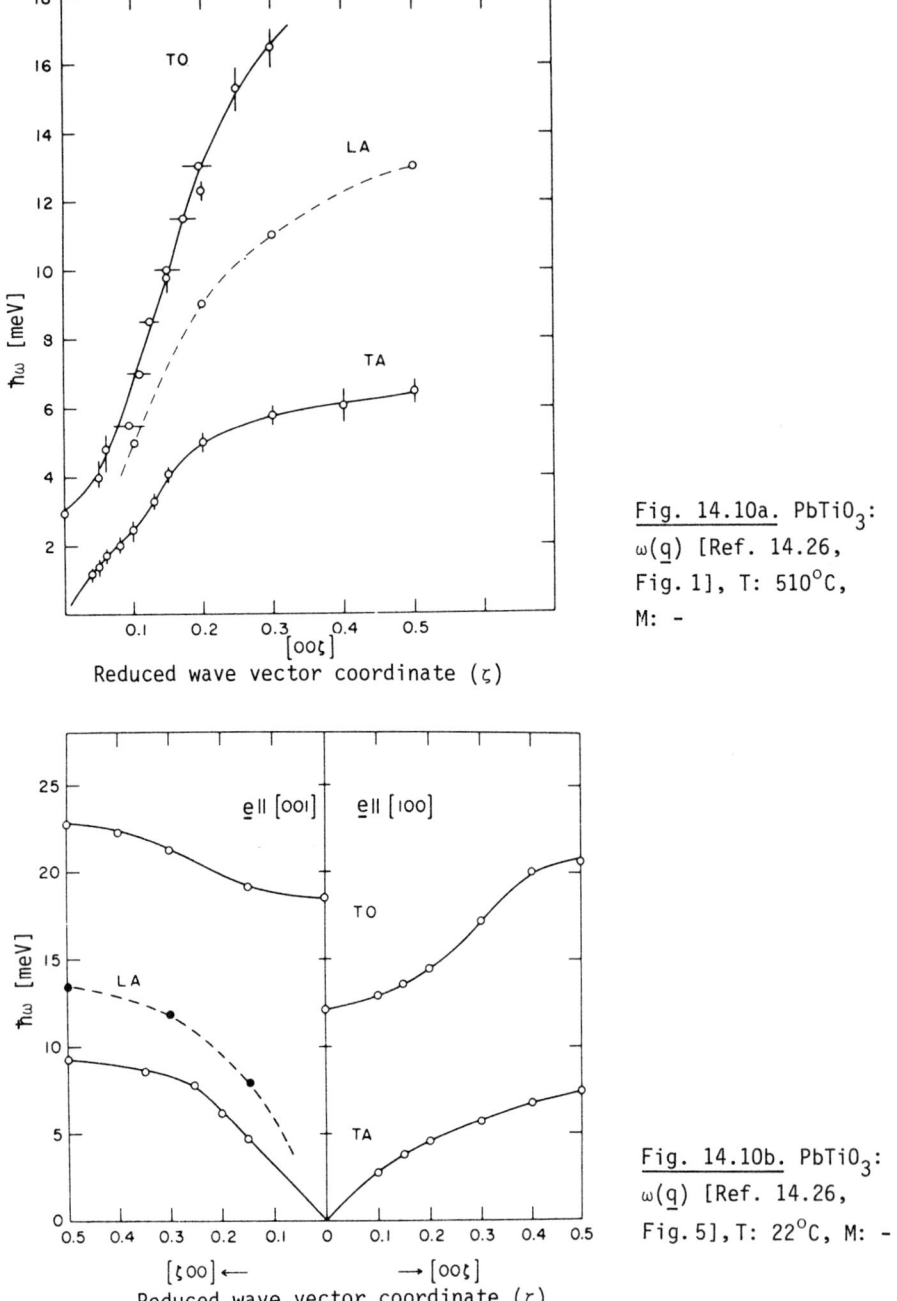

Fig. 14.10a. PbTiO$_3$: $\omega(\underline{q})$ [Ref. 14.26, Fig. 1], T: 510°C, M: -

Fig. 14.10b. PbTiO$_3$: $\omega(\underline{q})$ [Ref. 14.26, Fig. 5], T: 22°C, M: -

CsNiF₃

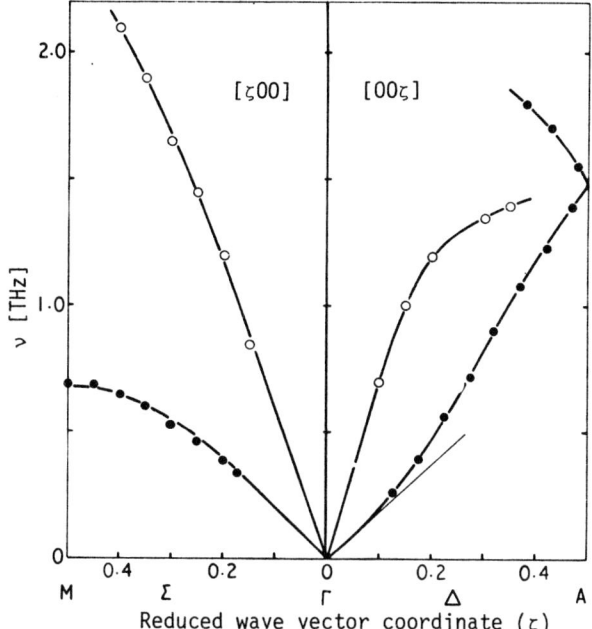

Fig. 14.11. $CsNiF_3$: $\omega(\underline{q})$ [Ref. 14.27, Fig. 2], T = 85 K, M: −

KCoF$_3$

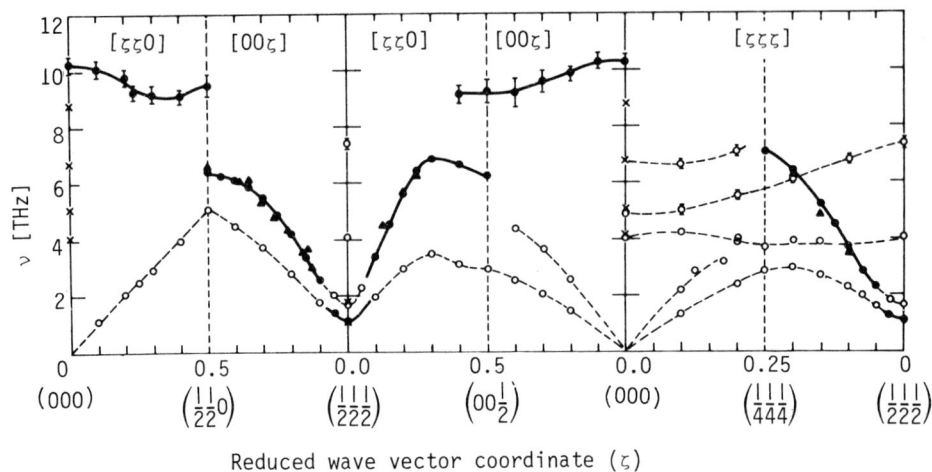

Fig. 14.12. KCoF$_3$: $\omega(\underline{q})$ [Ref. 14.28, Fig. 4], T = 22 K, M: -

CaCO₃

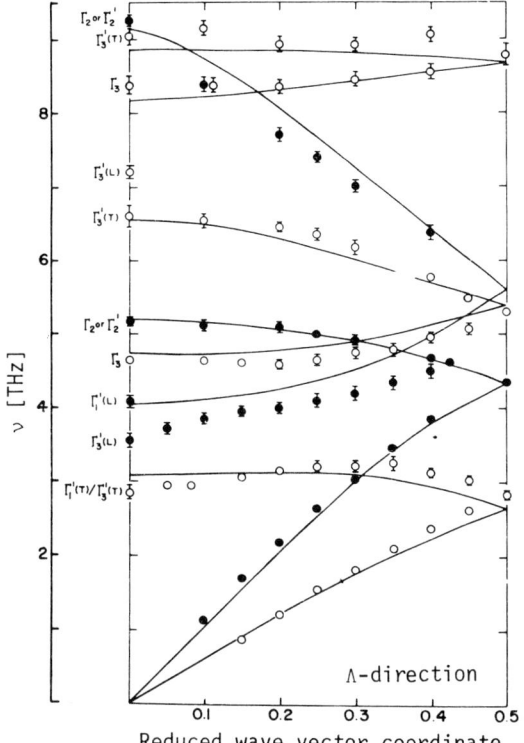

Fig. 14.13. $CaCO_3$: $\omega(\underline{q})$ [Ref. 14.29, Fig. 1], T = RT, M: 10P-SM

NaNO$_3$

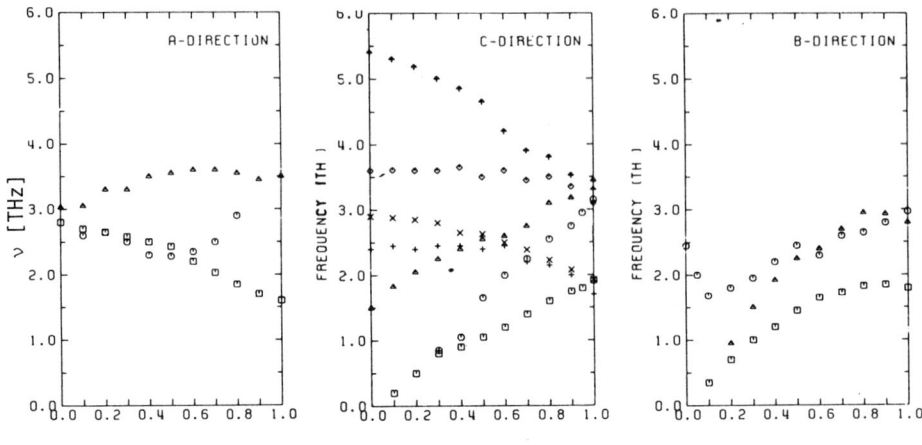

Fig. 14.14. NaNO$_3$: $\omega(\underline{q})$ [Ref. 14.30, Fig. 1], T = RT, M: -

KNO₃

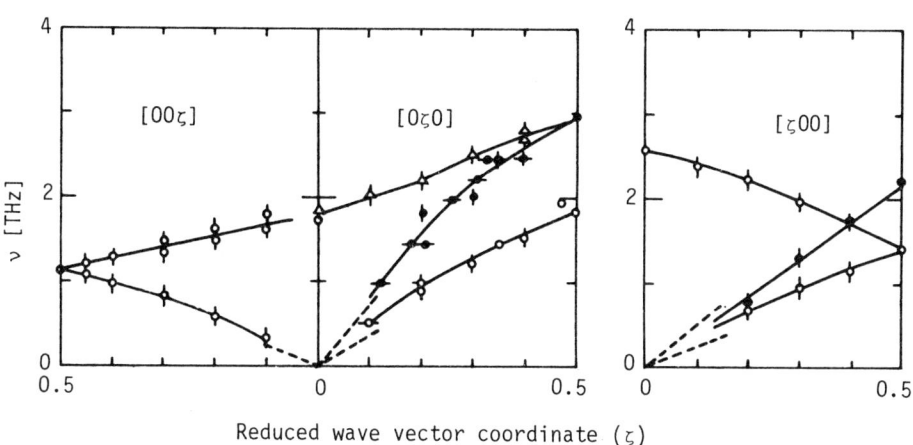

Fig. 14.15. KNO$_3$: $\omega(\underline{q})$ [Ref. 14.31, Fig. 1], T: RT, M: -, Lit. [14.32]

15. Layered Structure Crystals

Crystals with layer structures [15.31] are, at present, of particular interest. Their properties can very often be changed by intercalation, i.e., their structural anisotropy provides a short of chemical tool for a certain material "design" technique. The theoretical description is often done in terms of valence forces combined with axially symmetric forces. An important feature seems to be the structurally induced static dipoles [15.28,29] which lead to a quantitative understanding of the splitting between TO and LO branches. A full application of models used in cubic crystals should be very useful.

The following systems are treated:

Crystal	Figures showing		Crystal	Figures showing	
	Dispersion curve $\omega(\underline{q})$	Density of states $D(\omega)$		Dispersion curve $\omega(\underline{q})$	Density of states $D(\omega)$
Graphite	15.1a,b	15.1c	$CoCl_2$	15.8	
$NbSe_2$	15.2a,b		Bi_2Te_3	15.9a,b	
MoS_2	15.3a	15.3b	$SrGa_2$	15.10	
$SnSe_2$	15.4a,b		GaS	15.11a,b	15.11c
PbI_2	15.5		GaSe	15.12a	15.12b
$TiSe_2$	15.6a,b		GeS	15.13	
$FeCl_2$	15.7		TaS_2	15.14a,b	

Graphite

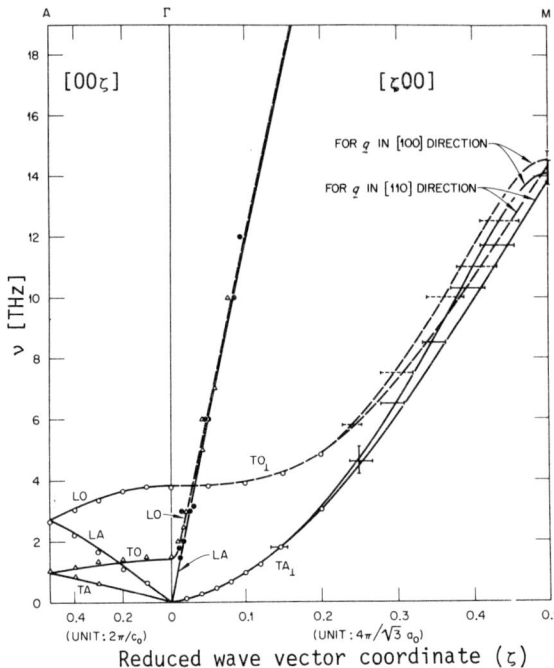

Fig. 15.1a. C (graphite): $\omega(q)$ [Ref. 15.1, Fig. 6], T = RT, M: 8P-ASM, Lit. [15.2-10]

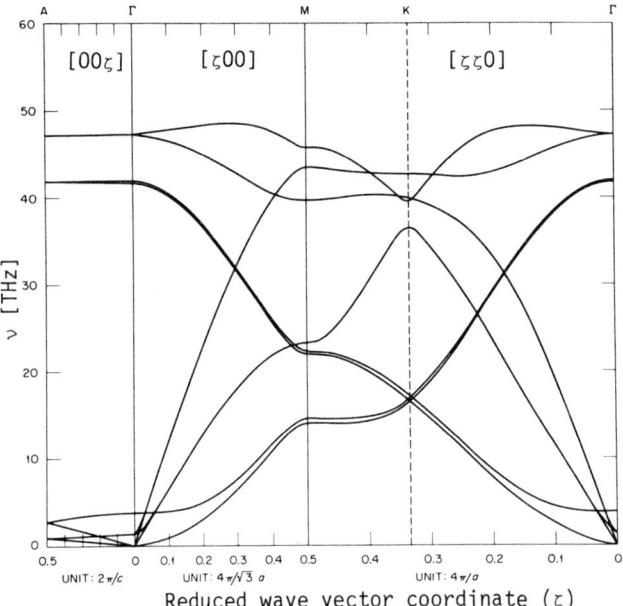

Fig. 15.1b. C (graphite): $\omega(q)$ [Ref. 15.1, Fig. 8], M: 8P-ASM Lit. [15.2-10]

Graphite

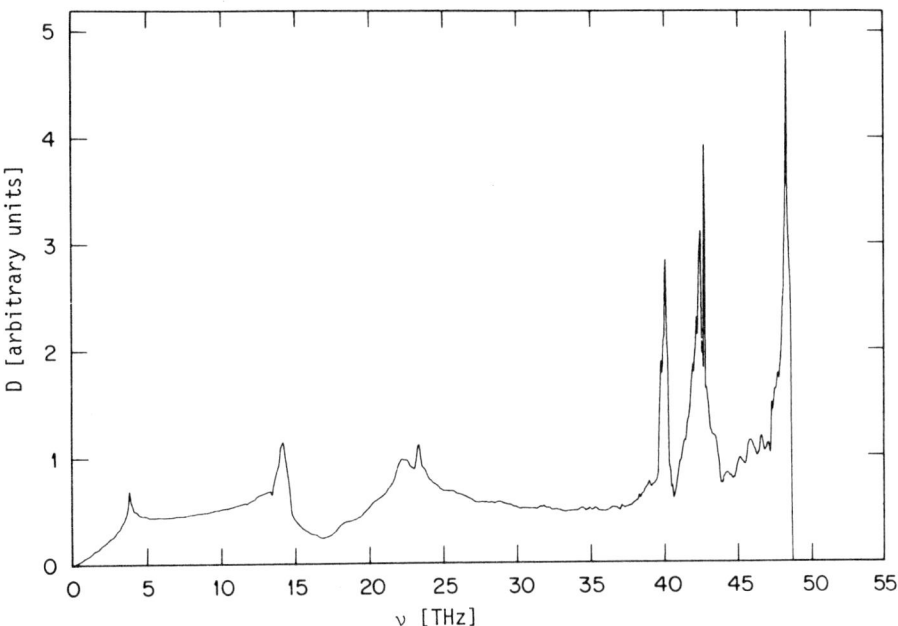

Fig. 15.1c. C (graphite): D(ω) [Ref.15.1, Fig.9], M: 8P-ASM, Lit.[15.2,4,5]

NbSe₂

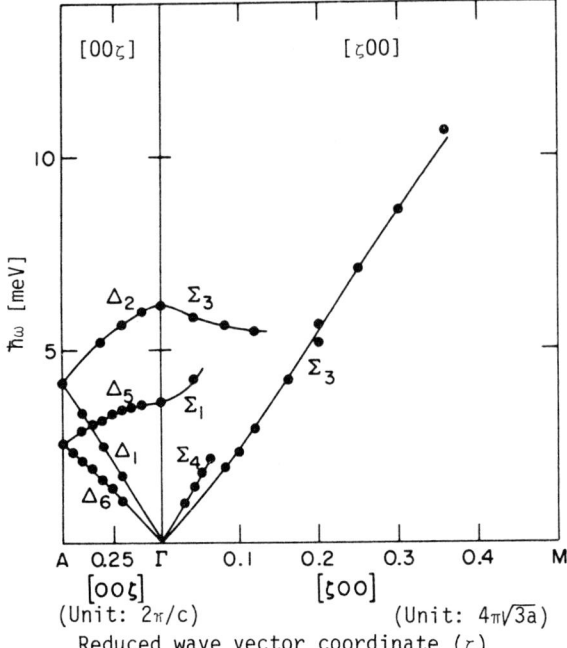

Fig. 15.2a. NbSe$_2$: $\omega(q)$ [Ref. 15.30, Fig. 12], T = 300 K, M: -, Lit. [15.11,12]

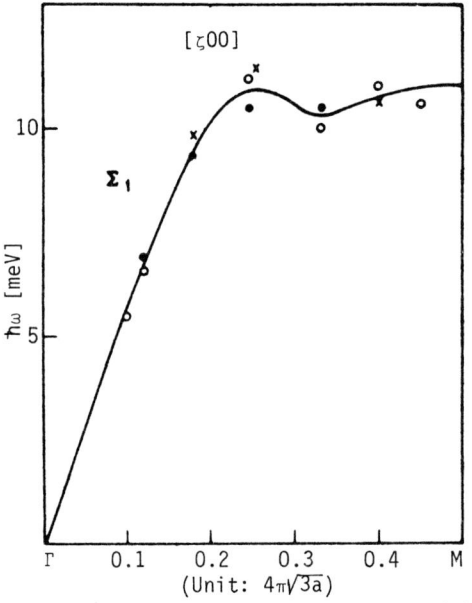

Fig. 15.2b. NbSe$_2$: $\omega(q)$ [Ref. 15.30, Fig. 13], T = 300 K, M: -, Lit. [15.11,12]

MoS$_2$

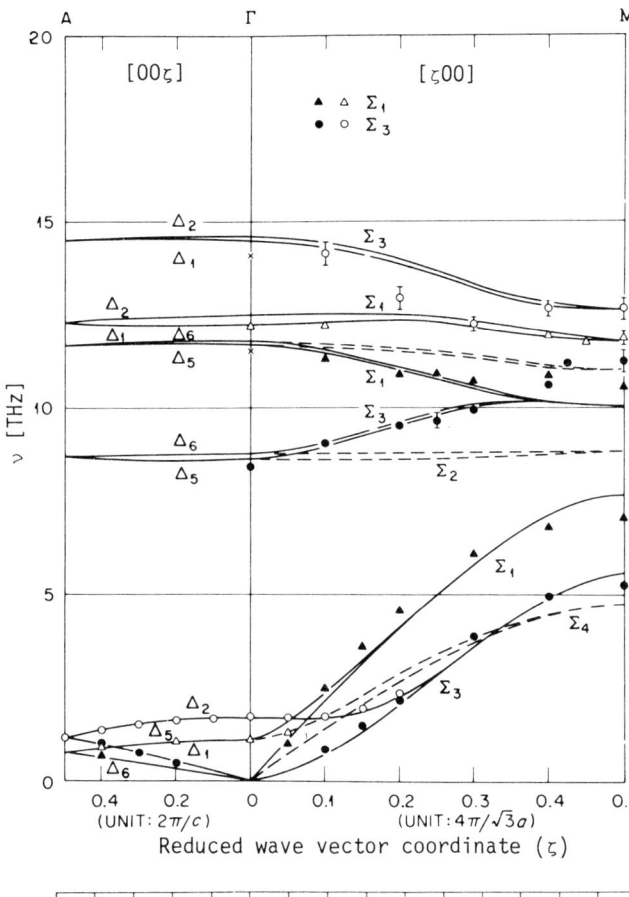

Fig. 15.3a. MoS$_2$: $\omega(q)$ [Ref. 15.13, Fig. 2], T = RT, M: 7P-FCM (VFF + AS), Lit. [15.12]

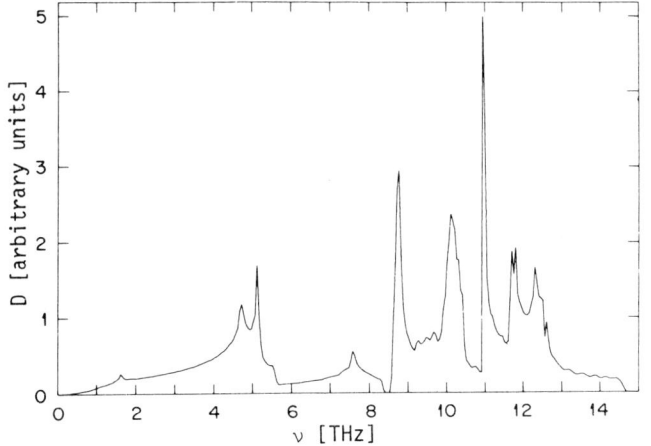

Fig. 15.3b. MoS$_2$: $D(\omega)$ [Ref. 15.13, Fig. 3], T = RT, M: 7P-FCM (VFF + AS), Lit. [15.12]

SnSe$_2$

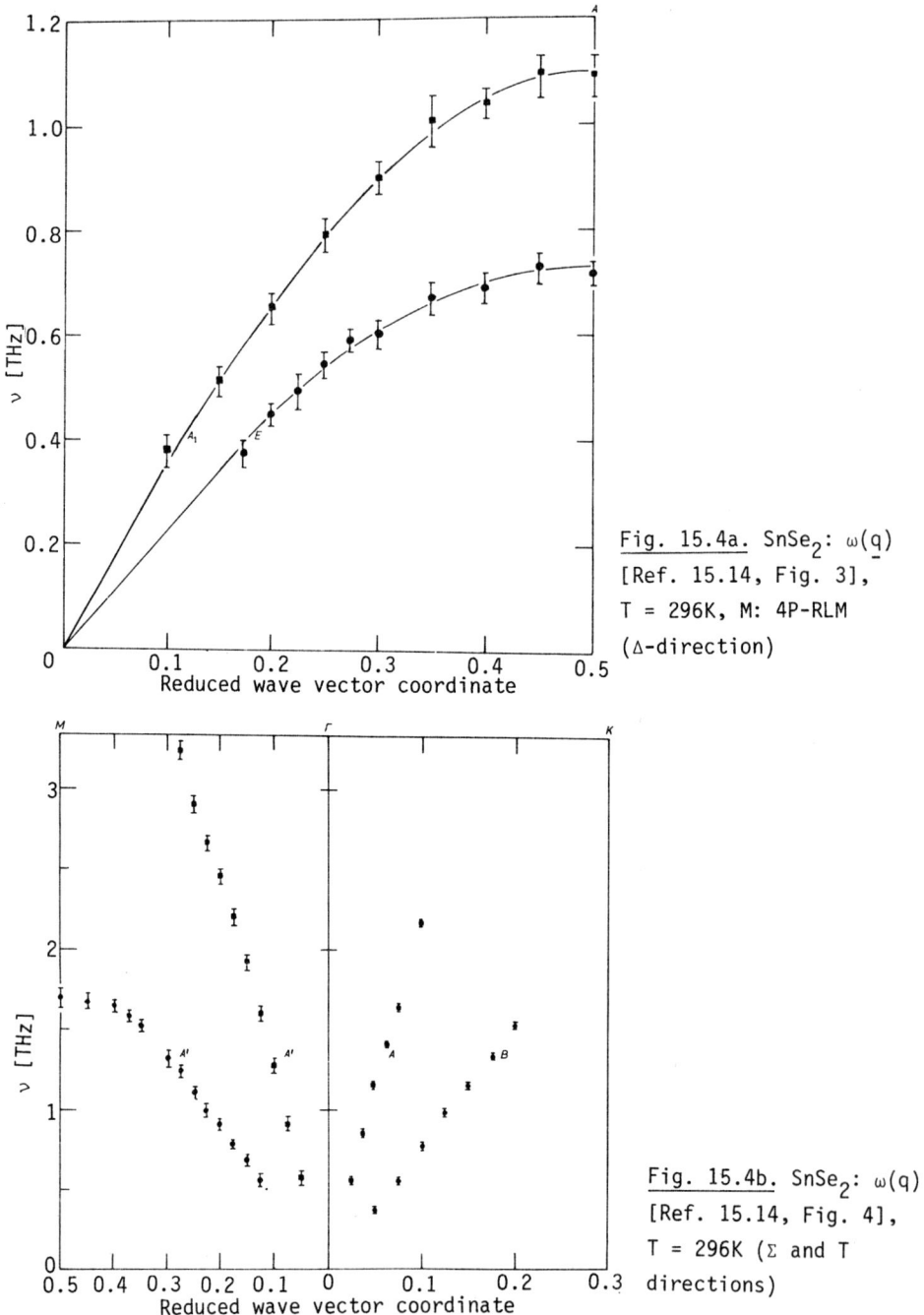

Fig. 15.4a. SnSe$_2$: $\omega(\underline{q})$ [Ref. 15.14, Fig. 3], T = 296K, M: 4P-RLM (Δ-direction)

Fig. 15.4b. SnSe$_2$: $\omega(\underline{q})$ [Ref. 15.14, Fig. 4], T = 296K (Σ and T directions)

PbI$_2$

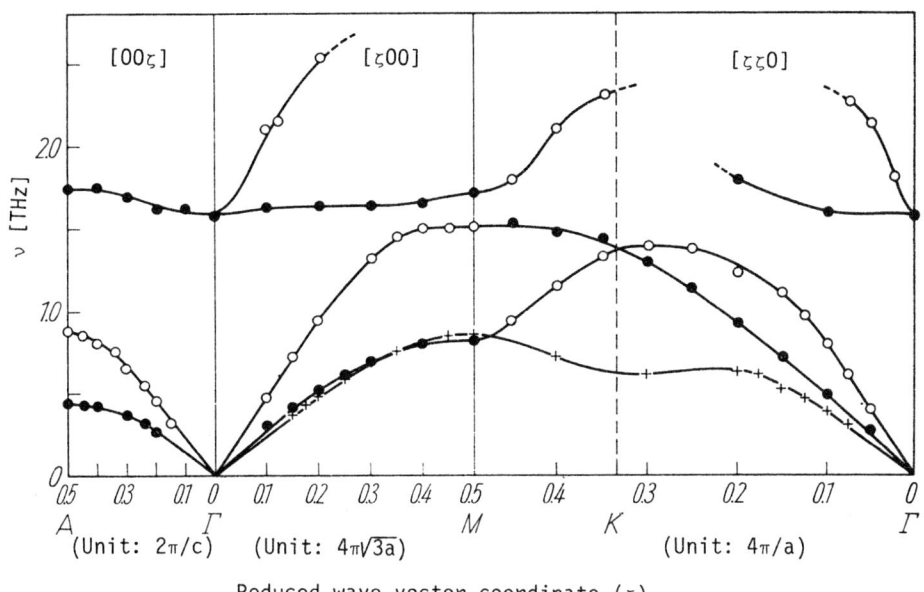

Fig. 15.5. PbI$_2$: ω(q) [Ref. 15.15, Fig. 4], T = RT, M: -, Lit. [15.12,29]

TiSe$_2$

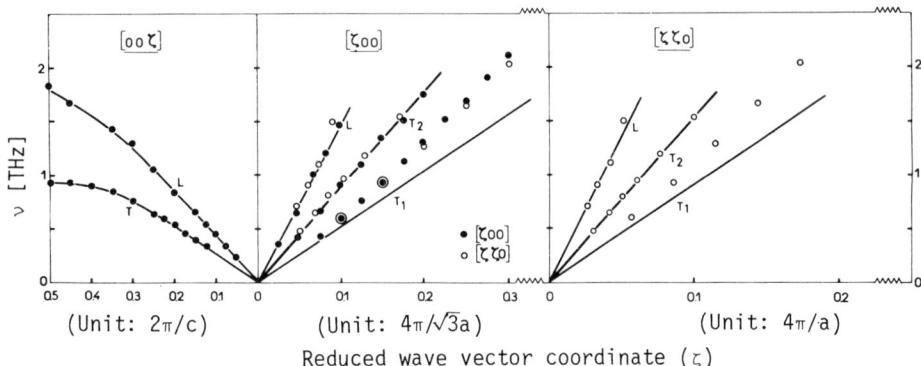

Fig. 15.6a. TiSe$_2$: $\omega(q)$ [Ref. 15.16, Fig. 2], T = RT, M: -

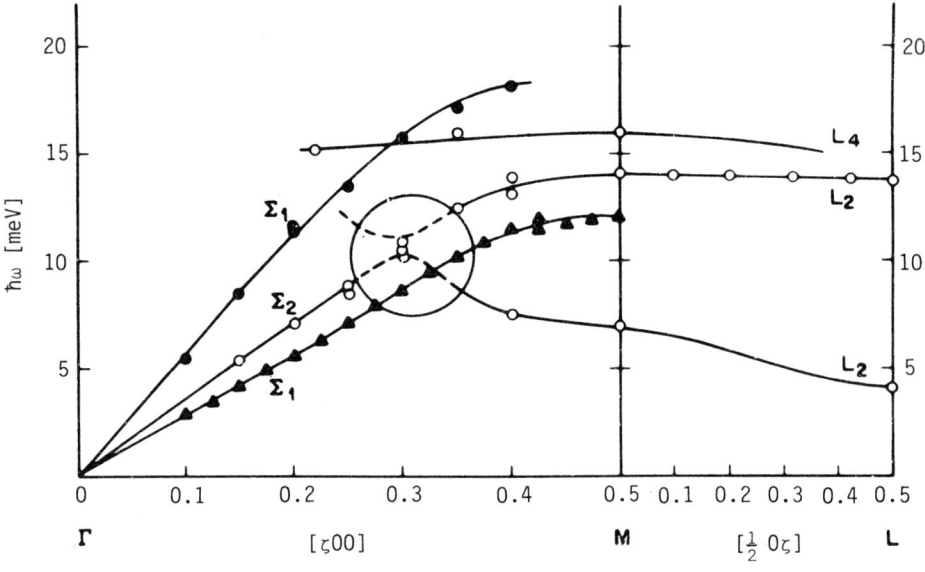

Fig. 15.6b. TiSe$_2$: $\omega(q)$ [Ref. 15.17, Fig. 4], T = RT, M: -

FeCl$_2$

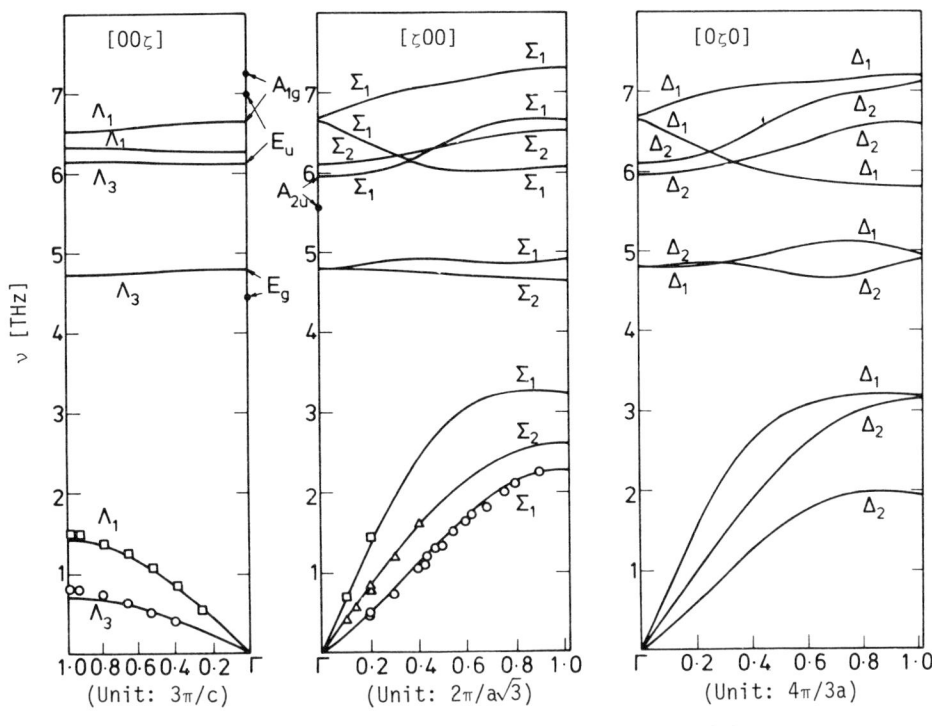

Fig. 15.7. FeCl$_2$: $\omega(\underline{q})$ [Ref. 15.18, Fig. 3b], measurements [15.19], T = RT, M: 8P-SM (static displaced shells included), Lit. [15.20]

CoCl₂

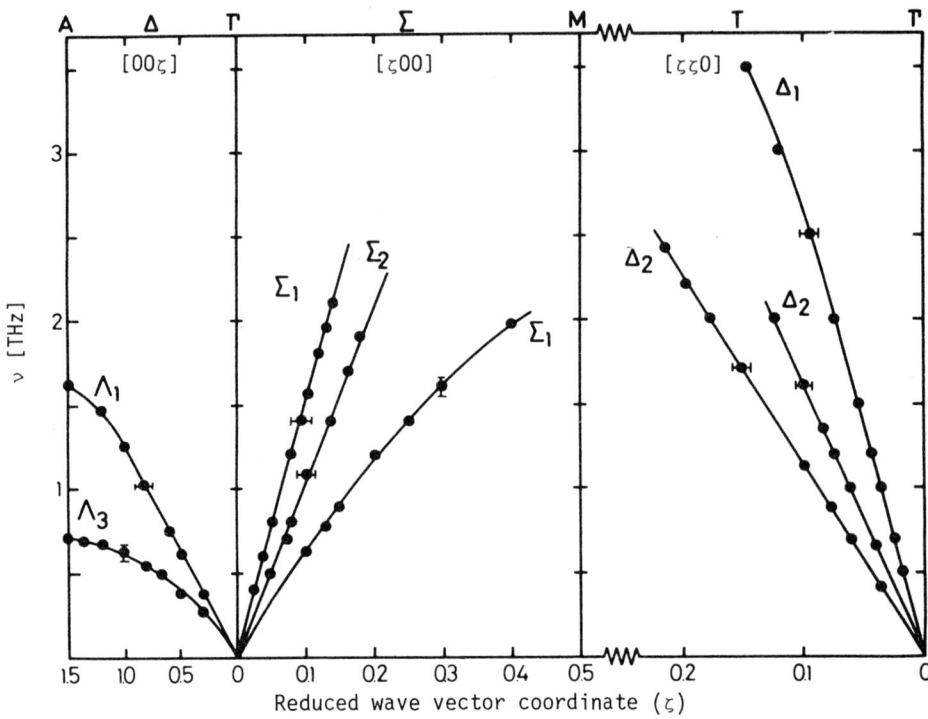

Fig. 15.8. CoCl$_2$: $\omega(\underline{q})$ [Ref. 15.21, Fig. 1], T = RT, Lit. [15.20], M: -

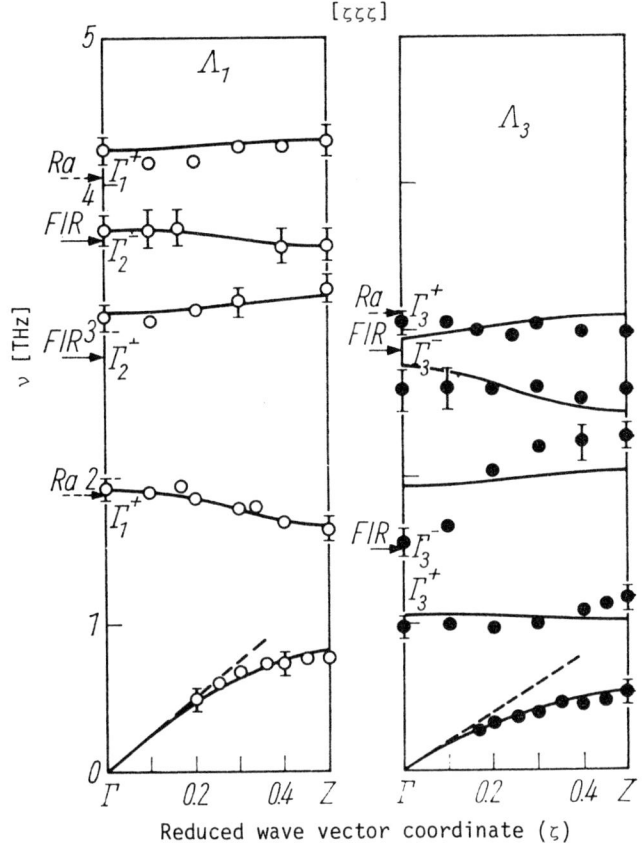

Fig. 15.9a. Bi_2Te_3: $\omega(\underline{q})$ [Ref. 15.22, Fig. 3], T = 77K, M: 12P-RLM modes propagating along the trigonal axis

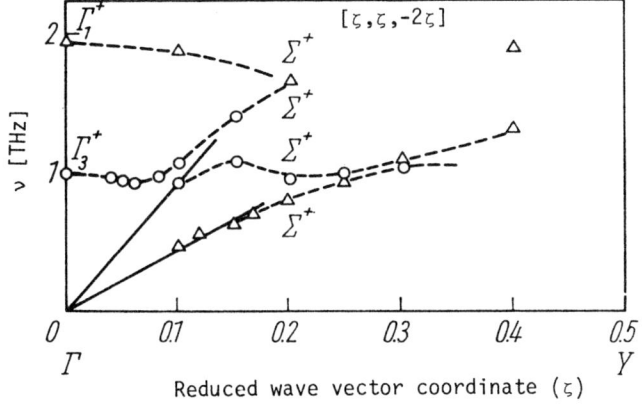

Fig. 15.9b. Bi_2Te_3: $\omega(\underline{q})$ [Ref. 15.22, Fig. 4], T = 77K, modes propagating along the bisectrix

SrGa$_2$

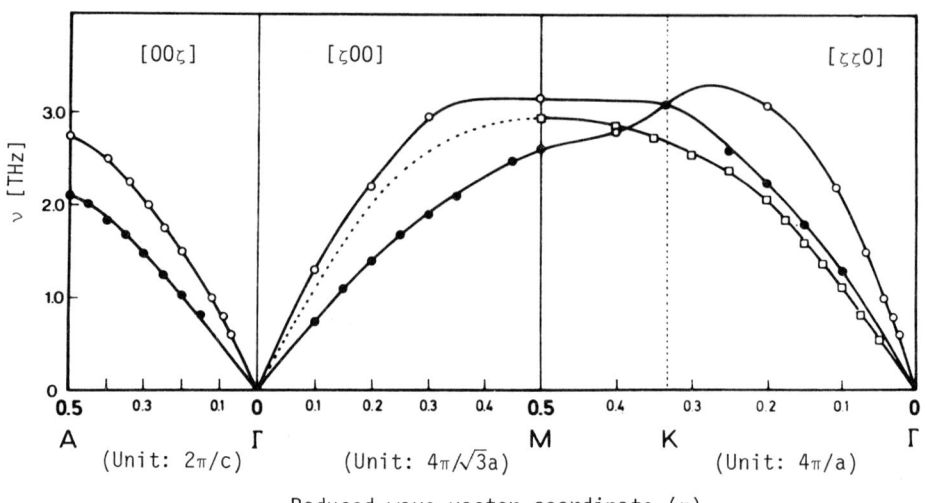

Fig. 15.10. SrGa$_2$: $\omega(\underline{q})$ [Ref. 15.23, Fig. 1], T = 20°C

GaS

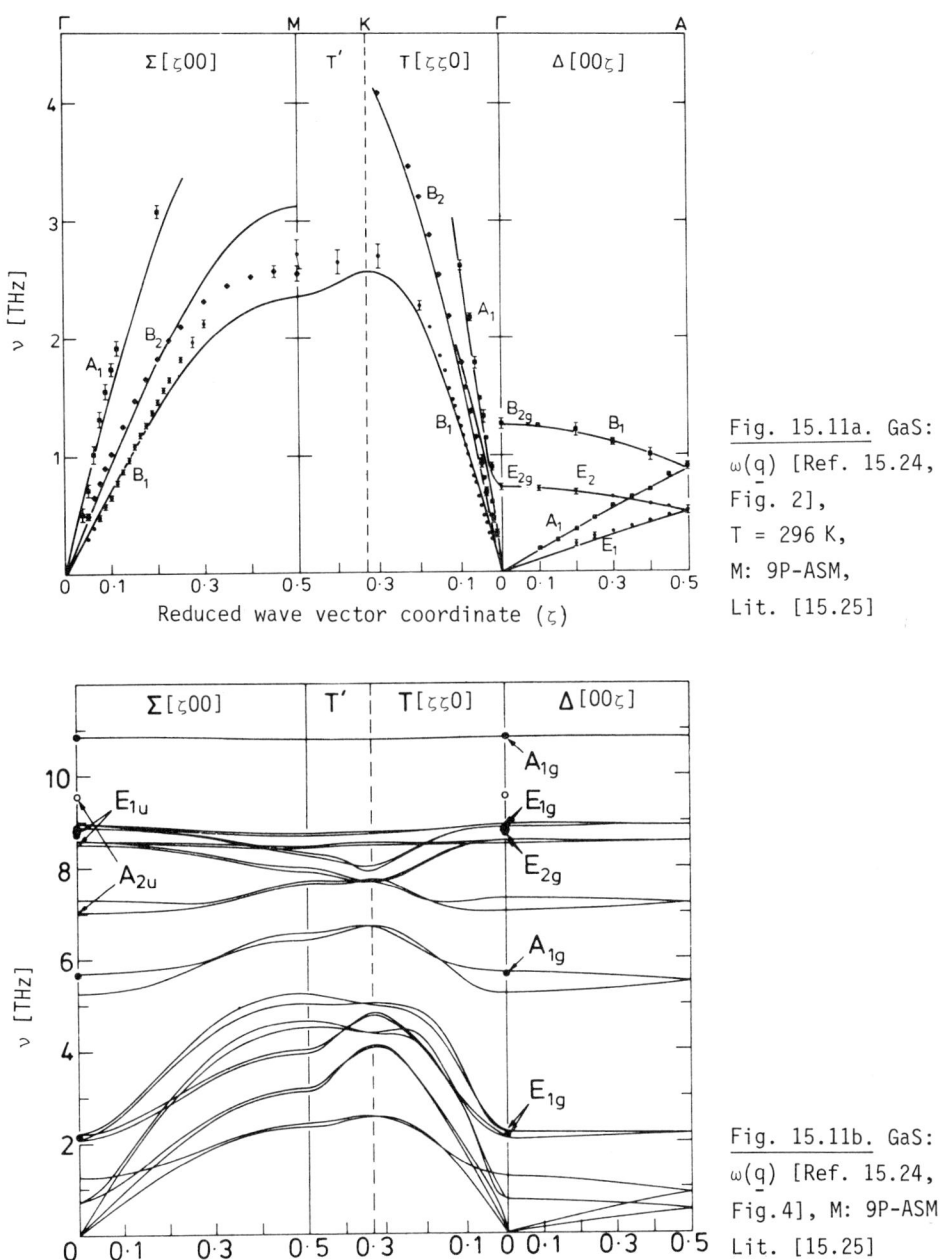

Fig. 15.11a. GaS: $\omega(\underline{q})$ [Ref. 15.24, Fig. 2], T = 296 K, M: 9P-ASM, Lit. [15.25]

Fig. 15.11b. GaS: $\omega(\underline{q})$ [Ref. 15.24, Fig. 4], M: 9P-ASM Lit. [15.25]

GaS

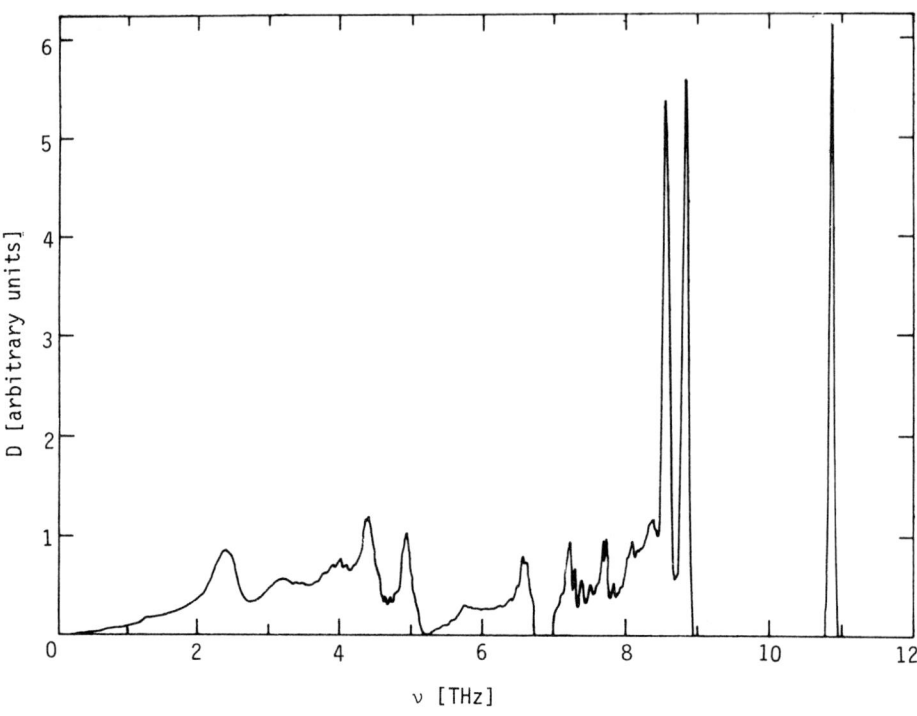

Fig. 15.11c. GaS: $D(\omega)$ [Ref. 15.24, Fig. 5], M: 9P-ASM

GaSe

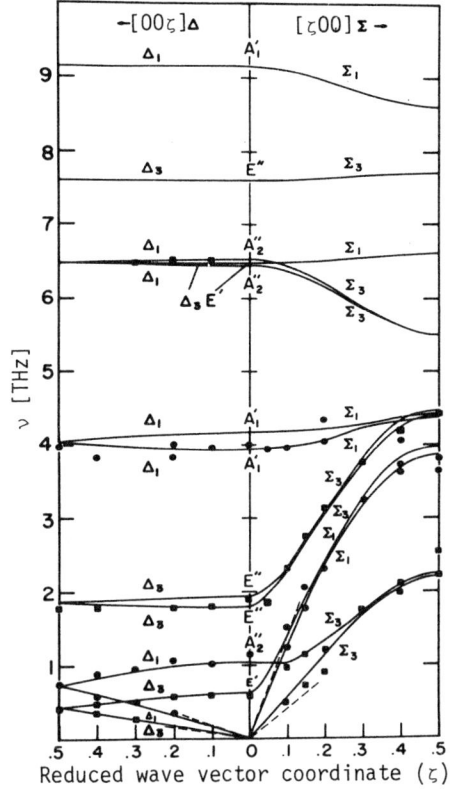

Fig. 15.12a. GaSe: $\omega(q)$ [Ref. 15.26, Fig. 3], T = 100 K, 9P-FCM

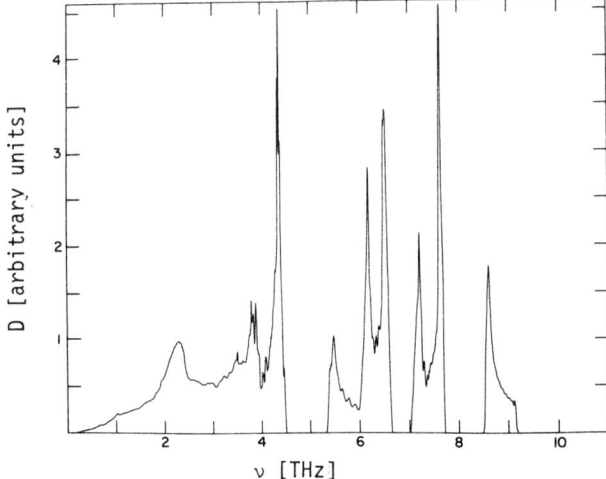

Fig. 15.12b. GaSe: $D(\omega)$ [Ref. 15.26, Fig. 5], T = 100 K, 9P-FCM

GeS

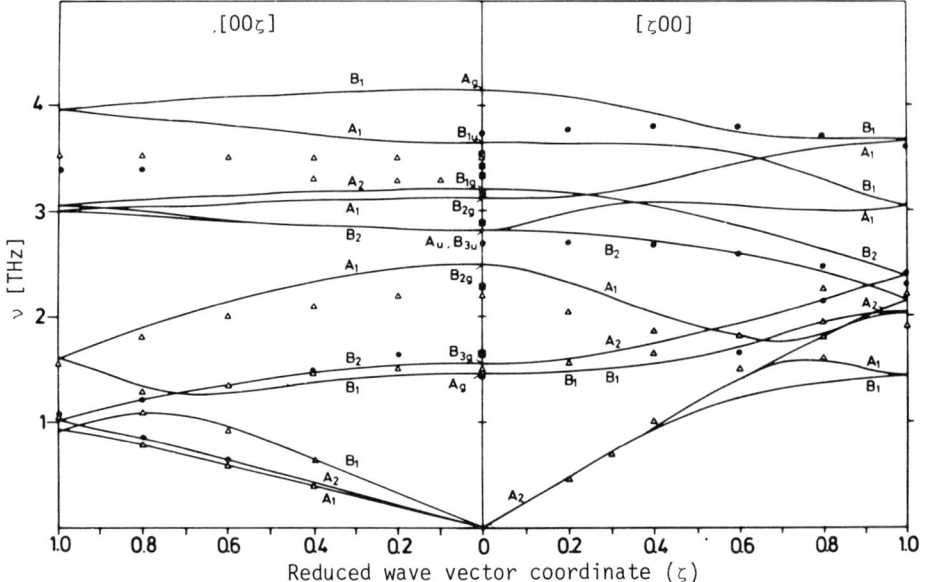

Fig. 15.13. GeS: ω(q) [Ref. 15.27, Fig. 1], T = RT, M: 4P-VFFM

TaSe$_2$

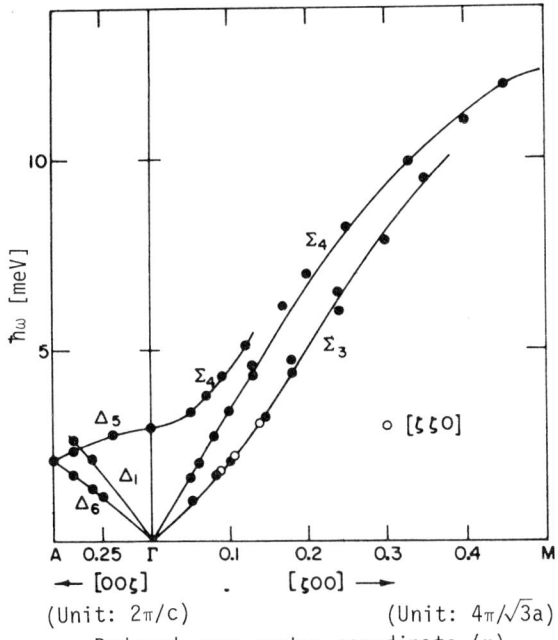

Fig. 15.14a. TaSe$_2$: $\omega(\underline{q})$ [Ref. 15.30, Fig. 11], T = 300 K, M: -

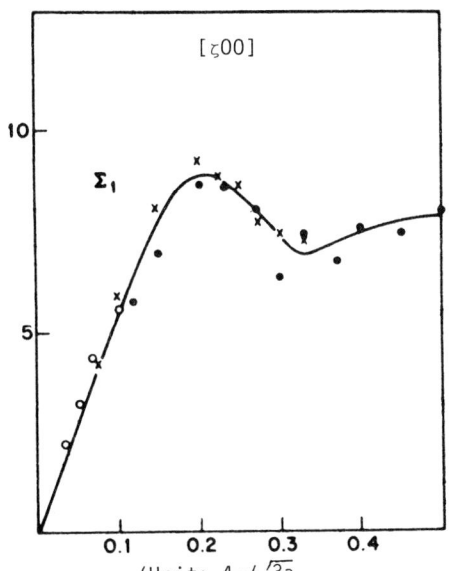

Fig. 15.14b. TaSe$_2$: $\omega(q)$ [Ref.15.30, Fig. 13], T = 300 K, M: -

16. Other Low-Symmetry Crystals

This chapter contains a rather heterogeneous group of crystals. They are all of low symmetry but it seems not worthwhile, at the present state of the art, to divide them up into smaller subgroups of (at most) two members. Useless to say that the lattice dynamics is very difficult but some successful analysis was done for Se and Te [16.7,13]. In the other cases, a more quantitative investigation is waiting to be carried out.

The following systems are treated:

Crystal	Figures showing		Crystal	Figures showing	
	Dispersion curve $\omega(\underline{q})$	Density of states $D(\omega)$		Dispersion curve $\omega(\underline{q})$	Density of states $D(\omega)$
Se	16.1a	16.1b	D_2O (ice, hex.)	16.8a	16.8b
Te	16.2a	16.2b	α-SiO_2 (quartz)	16.9	
$(SN)_x$	16.3		$MgAl_2O_4$	16.10	
S_8	16.4		K_2SeO_4	16.11	
α-As	16.5a	16.5b	KD_2PO_4	16.12	
NiS	16.6				
Cu_2O	16.7				

Se

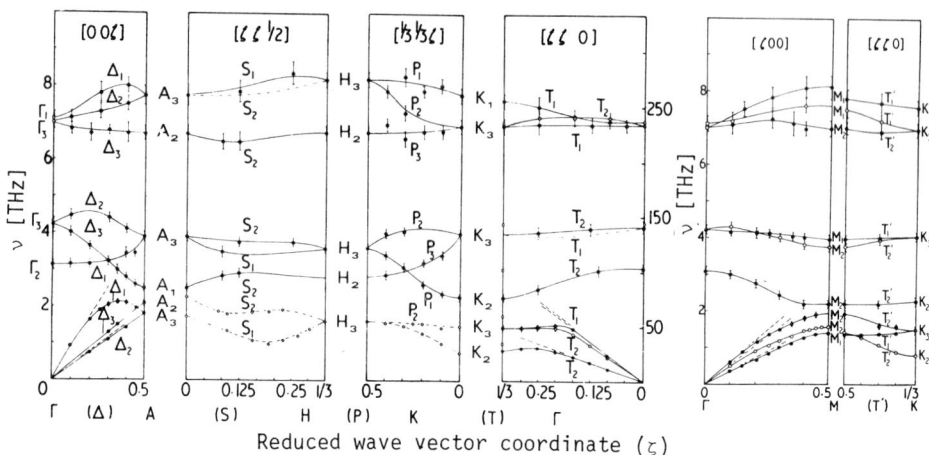

Fig. 16.1a. Se: $\omega(\underline{q})$ [Ref. 16.1, Fig.4], T = RT, M: -, rhombs [16.6], Lit. [16.2-7]

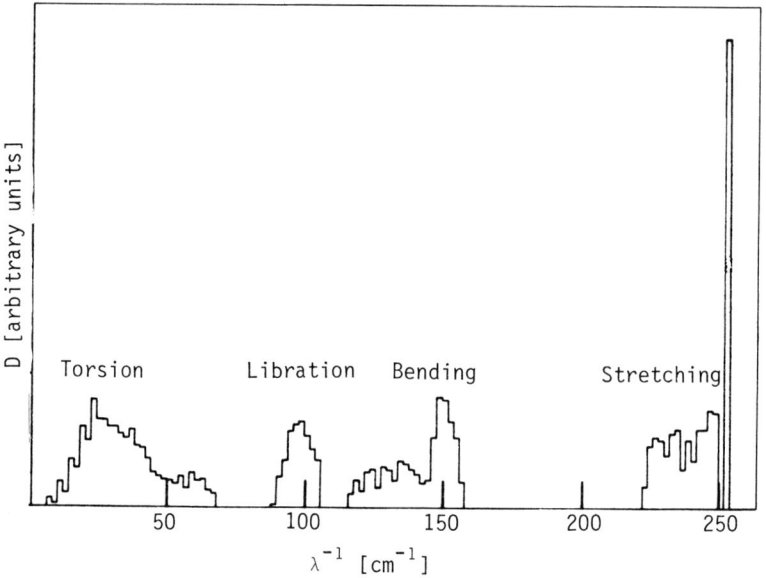

Fig. 16.1b. Se: $D(\omega)$ [Ref. 16.2, Fig.5], M: 9P-VFFM, Lit. [16.3]

Te

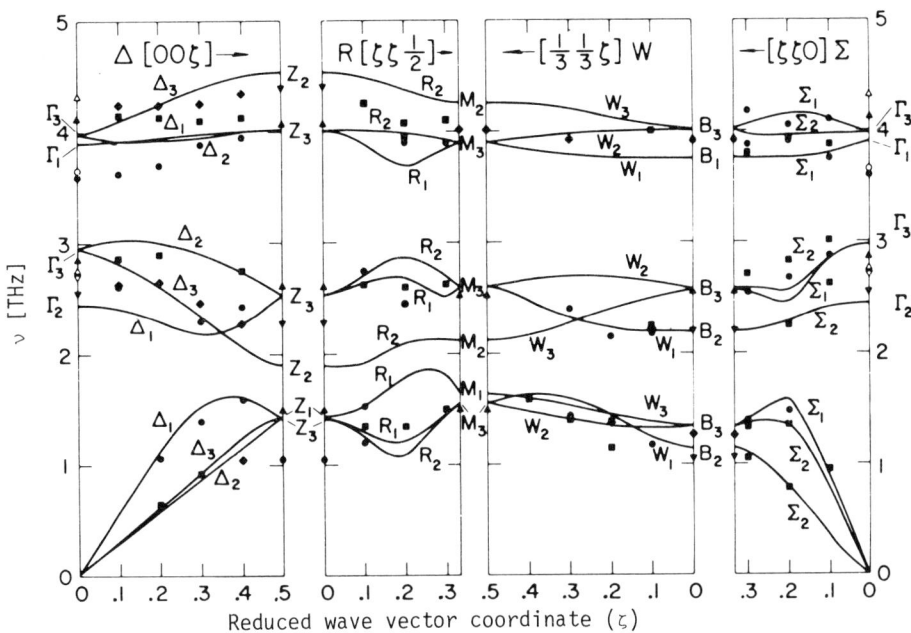

Fig. 16.2a. Te: $\omega(\underline{q})$ [Ref. 16.8, Fig. 3], T = 298 K, M: -, Lit. [16.2,9-13]

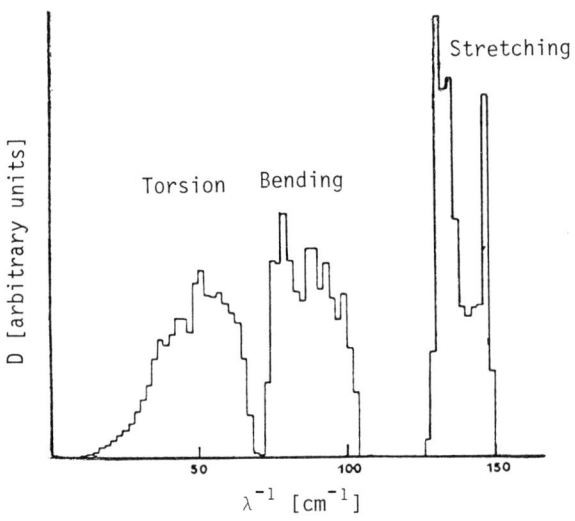

Fig. 16.2b. Te: $D(\omega)$ [Ref. 16.2, Fig. 6], M: 9P-VFFM, Lit. [16.8,10,12,30]

(SN)$_x$

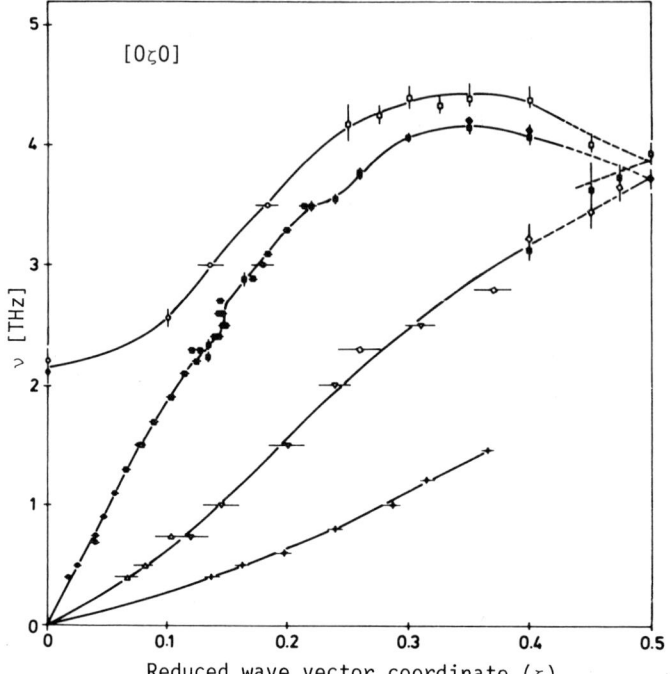

Fig. 16.3. (SN)$_x$: $\omega(q)$ [Ref. 16.14, Fig. 2], T = RT, M: -, Lit. [16.15]

S_8

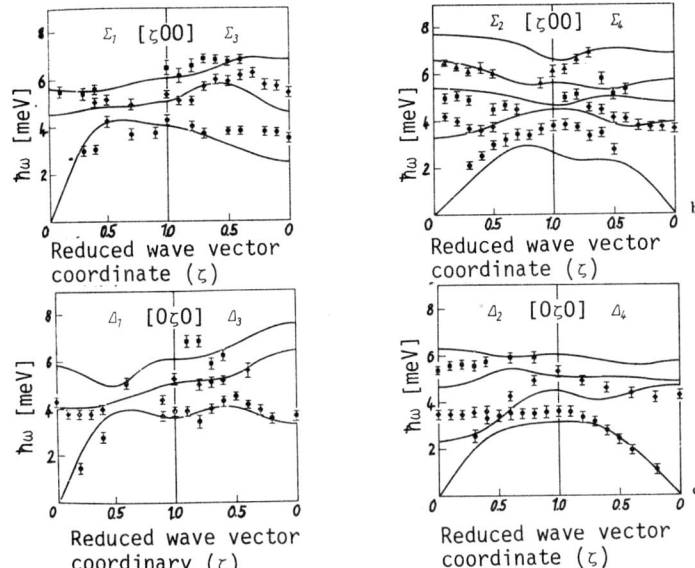

Fig. 16.4. α-S_8 ω($\underline{q}$) [Ref. 16.16, Fig. 2], T = RT, M: 5P-6-exp.-SM, measurements [16.29]

α-As

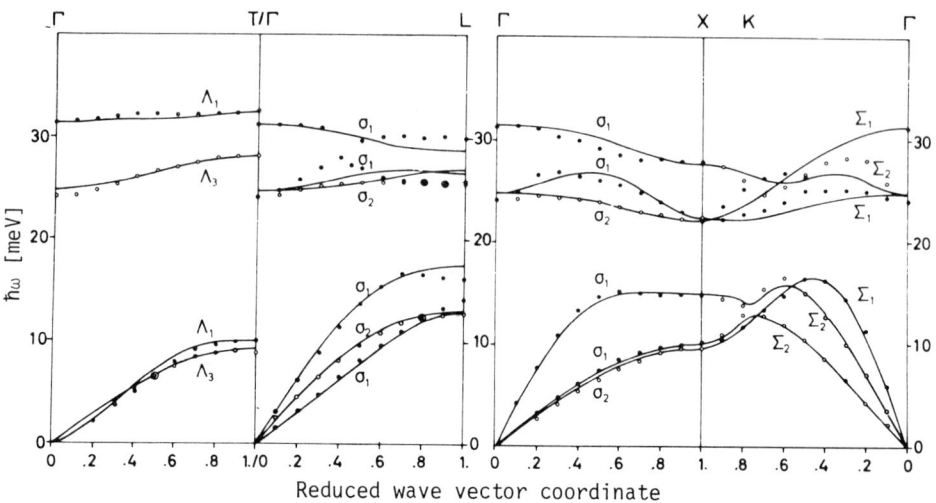

Fig. 16.5a. α-As: ω(q) [Ref. 16.17, Fig. 4], T = 297 K, M: 27P-FCM

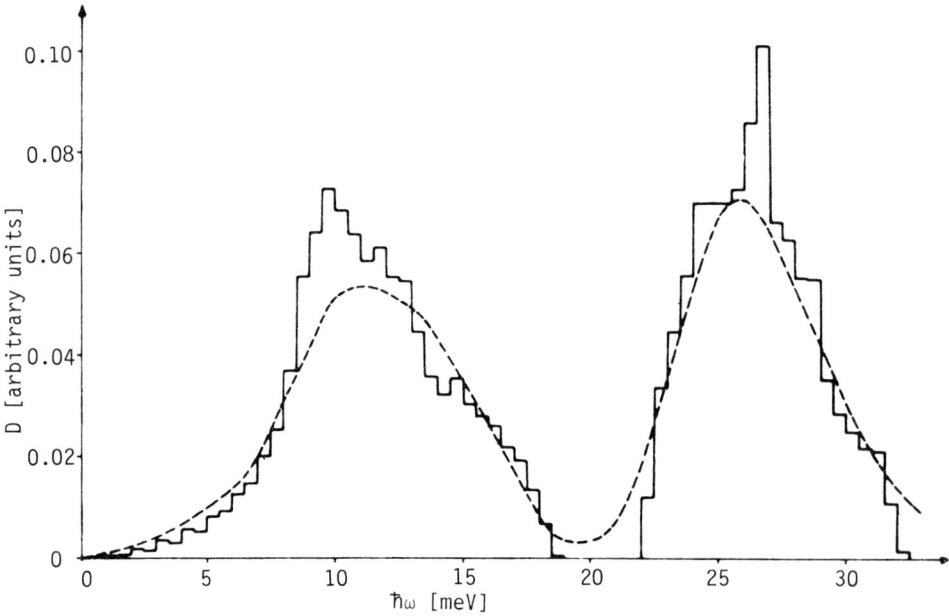

Fig. 16.5b. α-As: D(ω) [Ref. 16.17, Fig. 5], T = 297K, M: 27P-FCM

NiS

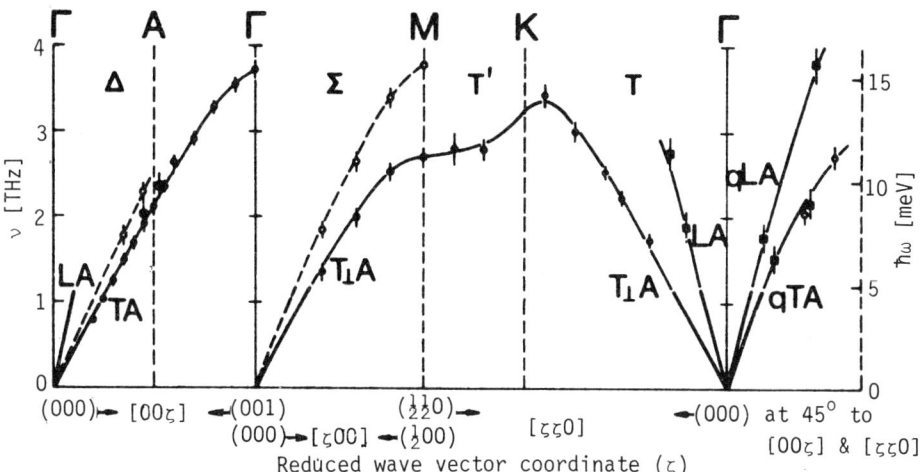

Fig. 16.6. NiS: $\omega(\underline{q})$ [Ref. 16.18, Fig. 3], T = 293 K (full symbols), 44 K (open symbols), M: -

Cu$_2$O

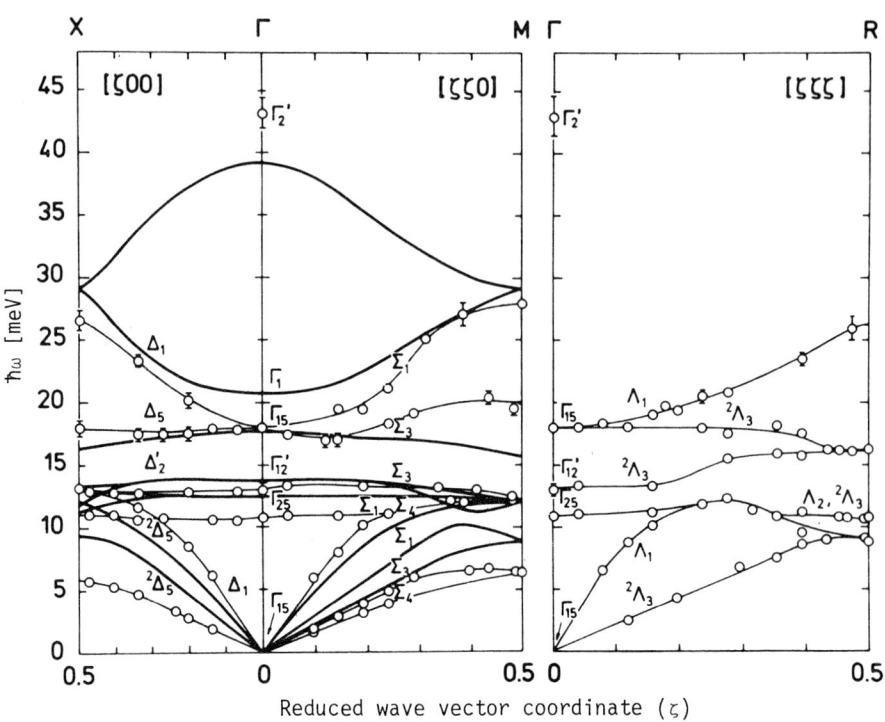

Fig. 16.7. Cu$_2$O: $\omega(\underline{q})$ [Ref. 16.19, Fig. 3], T = 20°C, M: 3P-RIM (heavy lines), —— (thin lines), Lit. [16.20]

D_2O (ice, hexagonal)

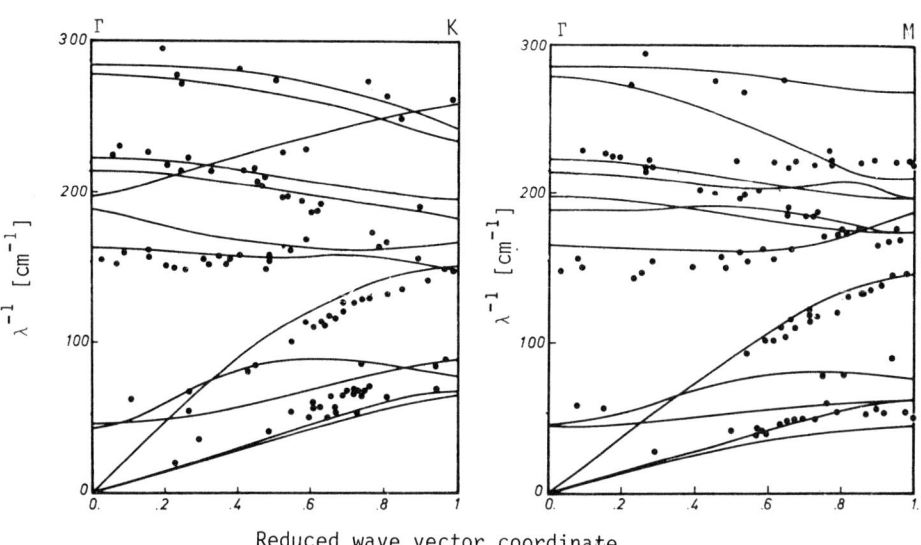

Fig. 16.8a. Ice (hexagonal) (D_2O): $\omega(\bar{q})$ [Ref. 16.32, Fig. 3], T = 90 K, M: 10P-VFFM, measured points [16.33]

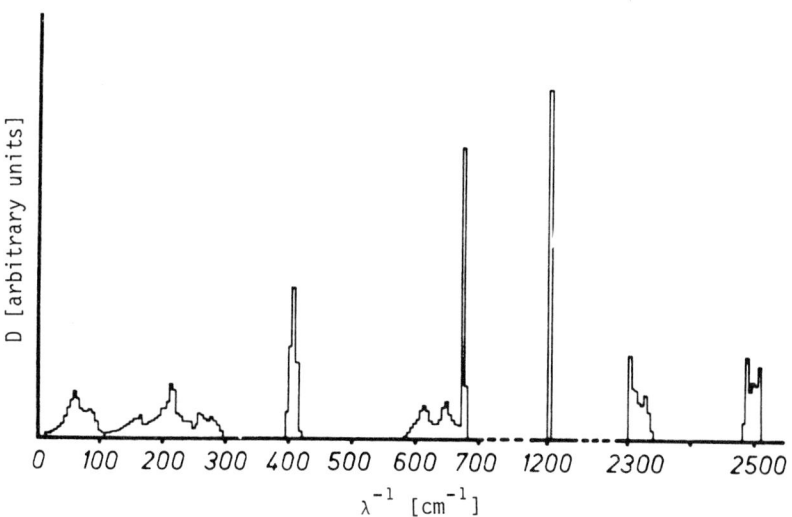

Fig. 16.8b. Ice (hexagonal) (D_2O): $D(\omega)$ [Ref. 16.32, Fig. 6], T = 90 K, M: 11P-VFFM, Lit. [16.21]

α-SiO₂ (quartz)

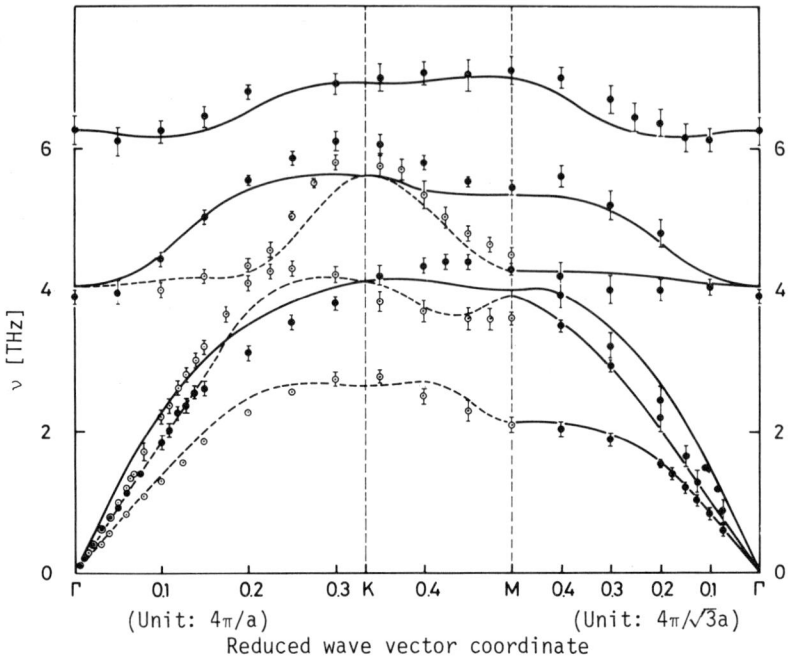

Fig. 16.9. α-quartz (α-SiO$_2$): ω(q) [Ref. 16.23], T = RT, M: 7P-FCM [16.22], Lit. [16.24-26]

$MgAl_2O_4$

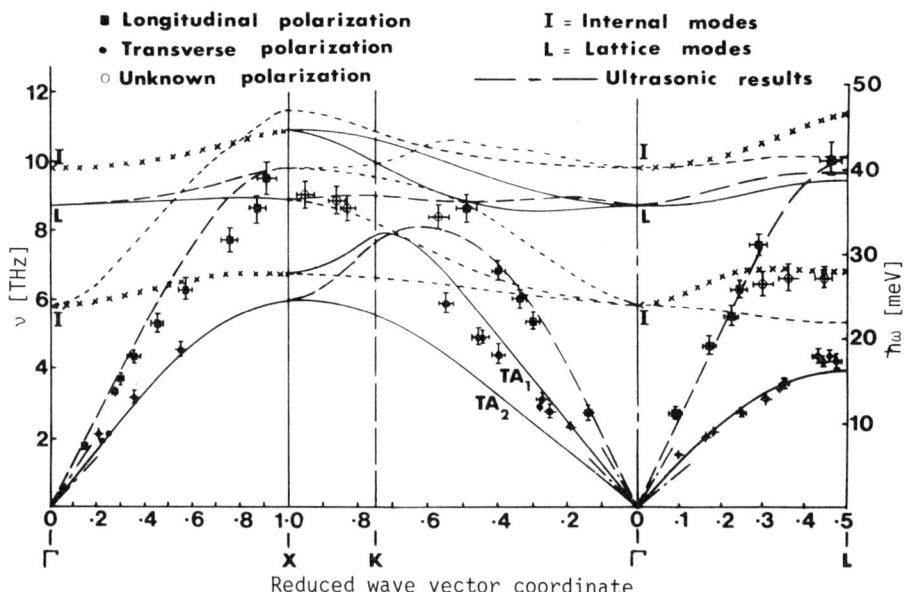

Fig. 16.10. $MgAl_2O_4$: $\omega(\underline{q})$ [Ref. 16.27, Fig. 3], T = RT, M: -

K$_2$SeO$_4$

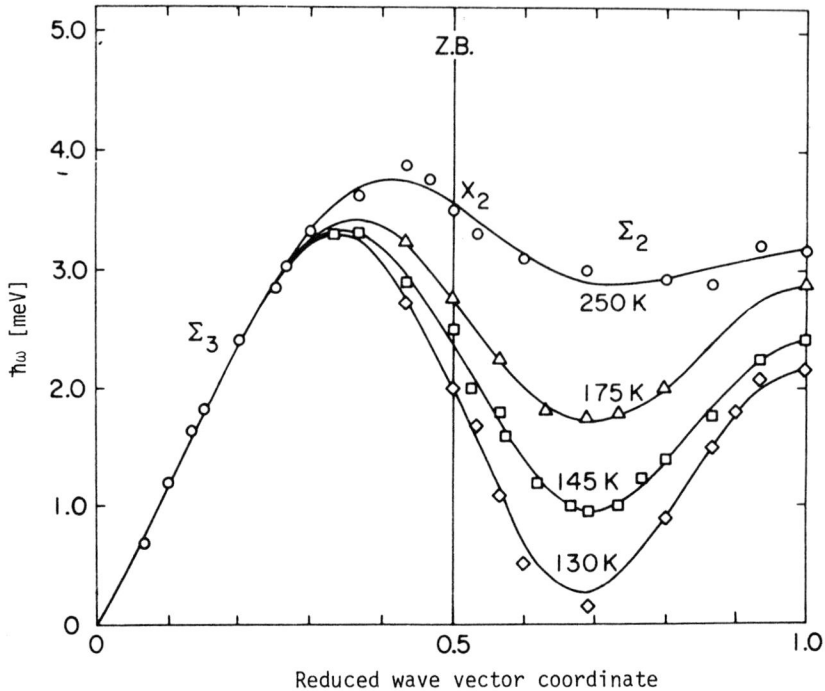

Fig. 16.11. K$_2$SeO$_4$: $\omega(\underline{q})$ [Ref. 16.28, Fig. 9] plotted in an extended zone; Z.B. indicates the zone boundary of the original zone, M: FCM

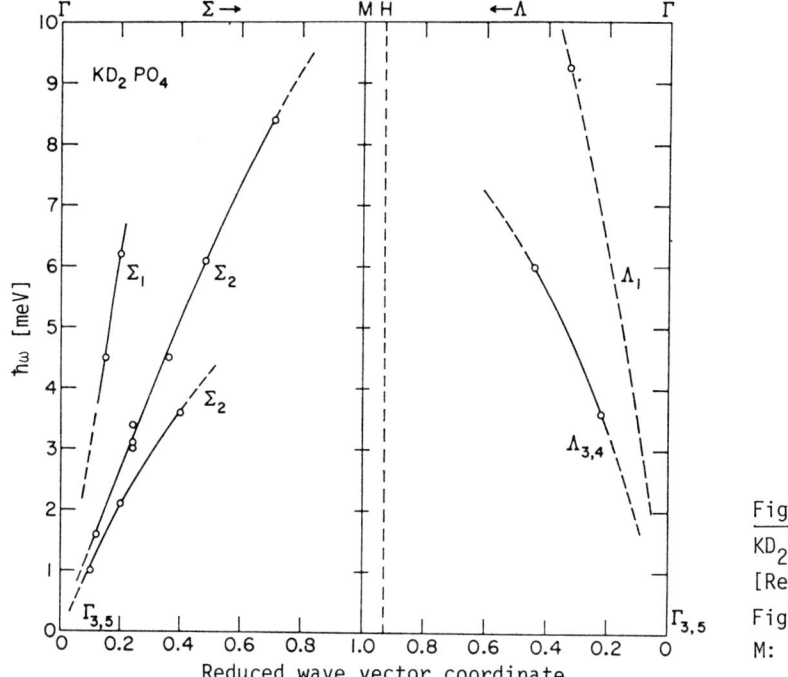

Fig. 16.12. KD_2PO_4: $\omega(\underline{q})$ [Ref.16.31, Fig.5], T = RT, M: –

17. Molecular Crystals

The definition of molecular crystals by the separation of intra- and intermolecular vibrations is sometimes difficult. The concept of this separation is certainly useful and provides a starting point for the theoretical analysis. The field is rapidly growing and more basic treatments are required in order to understand the delicate playing together of both types of modes in these crystals. Two molecular crystals, $NaNO_3$ and KNO_3, may be found in Chap. 14 among the ABO_3 structures.

The following systems are treated:

Crystal	Figures showing	
	Dispersion curve $\omega(\underline{q})$	Density of states $D(\omega)$
α-N_2	17.1	
I_2	17.2	
CO_2	17.3	
NH_3		17.4
D_2O_2	17.5	
NaO_2	17.6	
KN_3	17.7	
SnI_4	17.8	
$NaNO_2$	17.9	

α-N₂

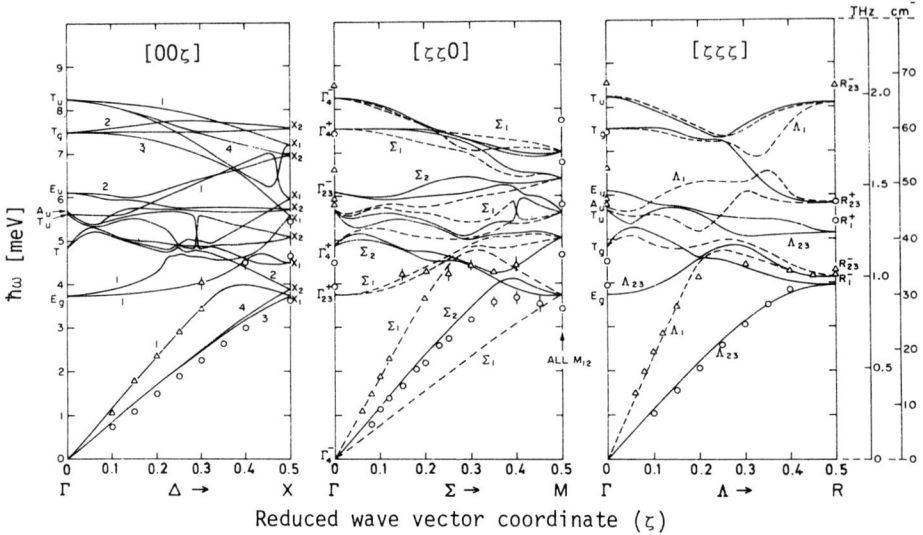

Fig. 17.1. α-N$_2$: ω(q) [Ref. 17.1, Fig. 5], T = 15 K, M: 3P-6/12-pot. including electrostatic quadrupol interaction

I_2

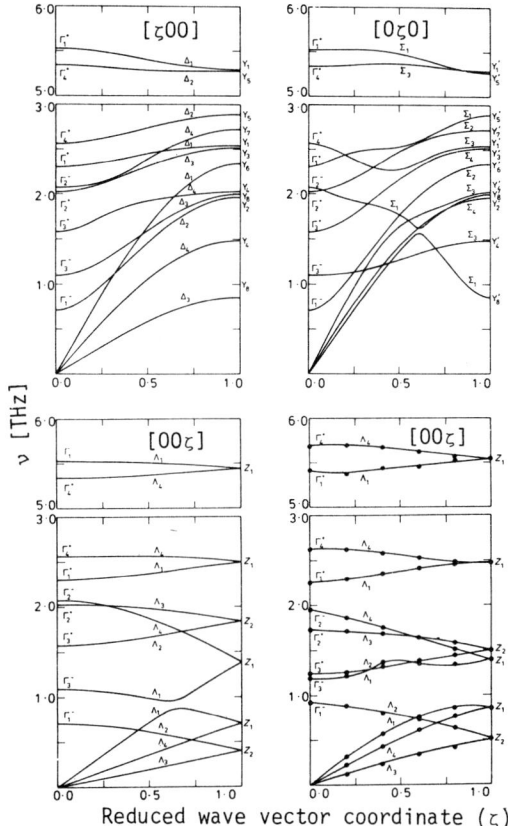

Fig. 17.2. I_2: $\omega(q)$ [Ref.17.2, Fig. 3], T = 77 K, measurements [17.3], M: 8P-BCM

CO_2

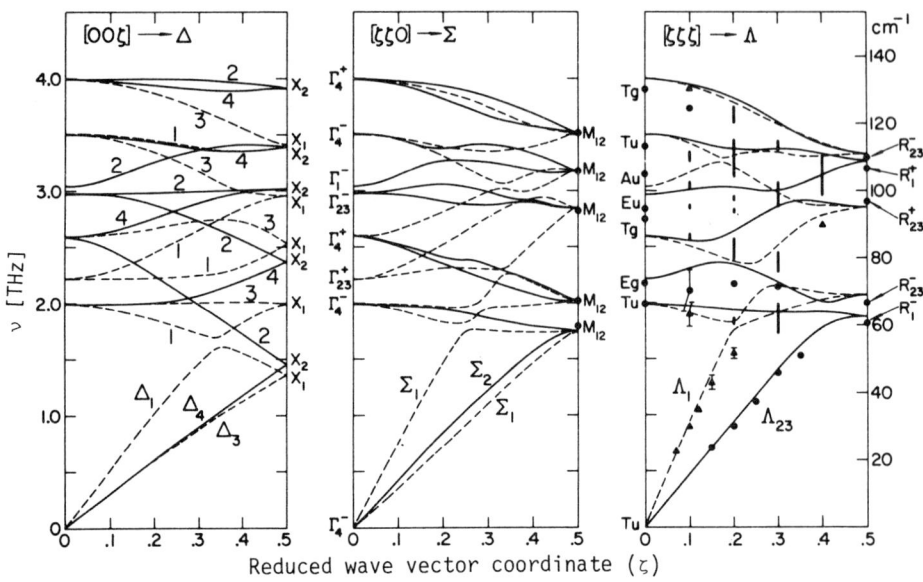

Fig. 17.3. CO_2: $\omega(\underline{q})$ [Ref. 17.4., Fig. 3], T = 95 K, M: 8P-RIM

NH₃

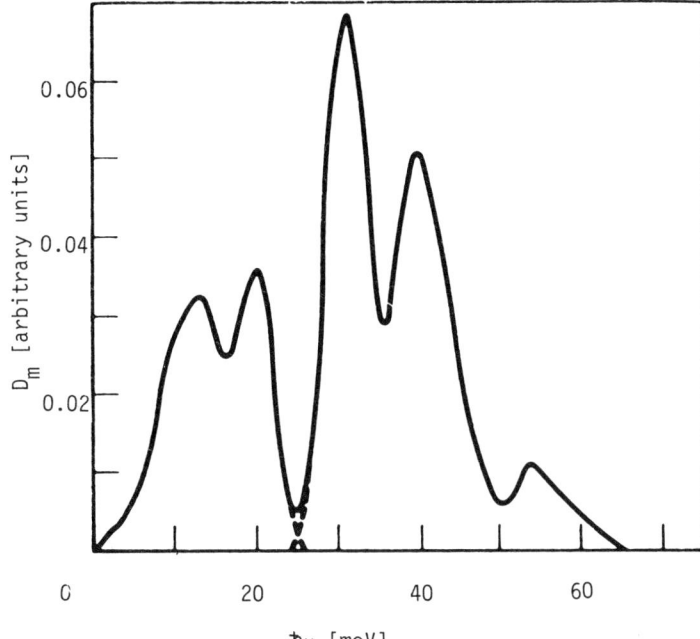

Fig. 17.4. NH$_3$: D(ω) [Ref.17.5, Fig.3], T = 106 K

D_2O_2

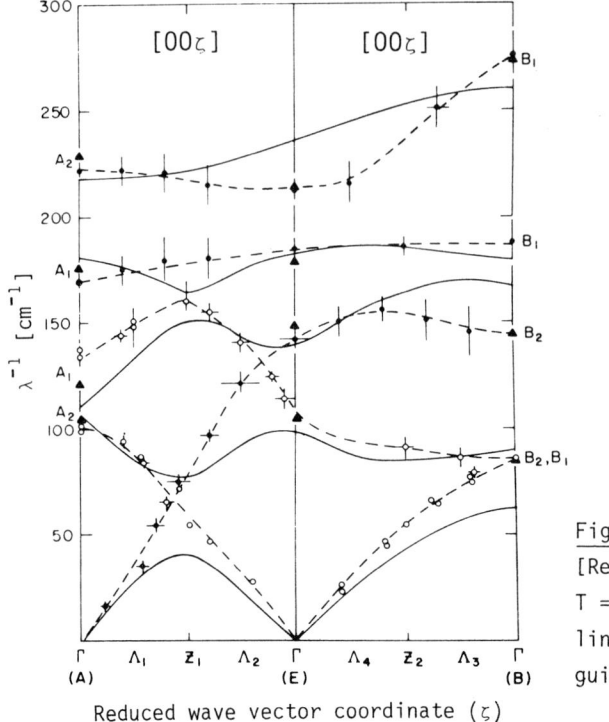

Fig. 17.5. D_2O_2: $\omega(\underline{q})$ [Ref. 17.6, Fig.1], T = RT, M: VFFM (full lines), dashed curves: guide lines to the eye

NaO₂

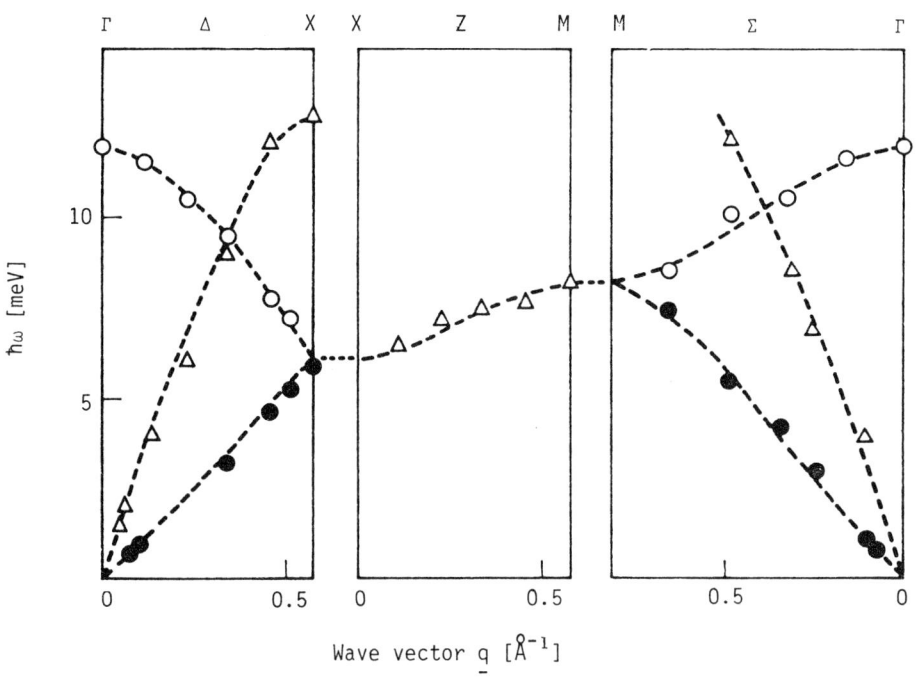

Fig. 17.6. NaO_2: $\omega(\underline{q})$ [Ref. 17.7, Fig.6], T = 225 K, M: -

KN₃

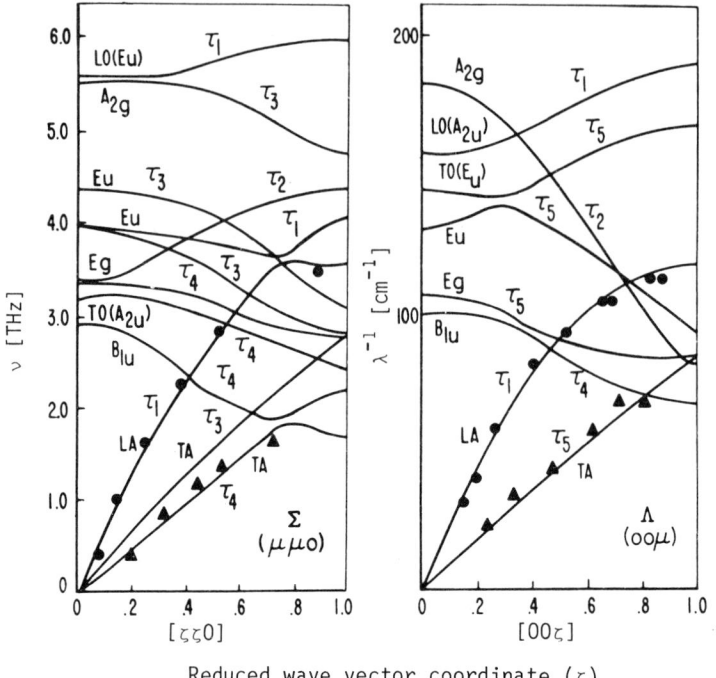

Fig. 17.7. KN$_3$: $\omega(q)$ [Ref. 17.8, Fig. 3], T = RT, M: 2P - pot.-M including Coulomb interactions

SnI$_4$

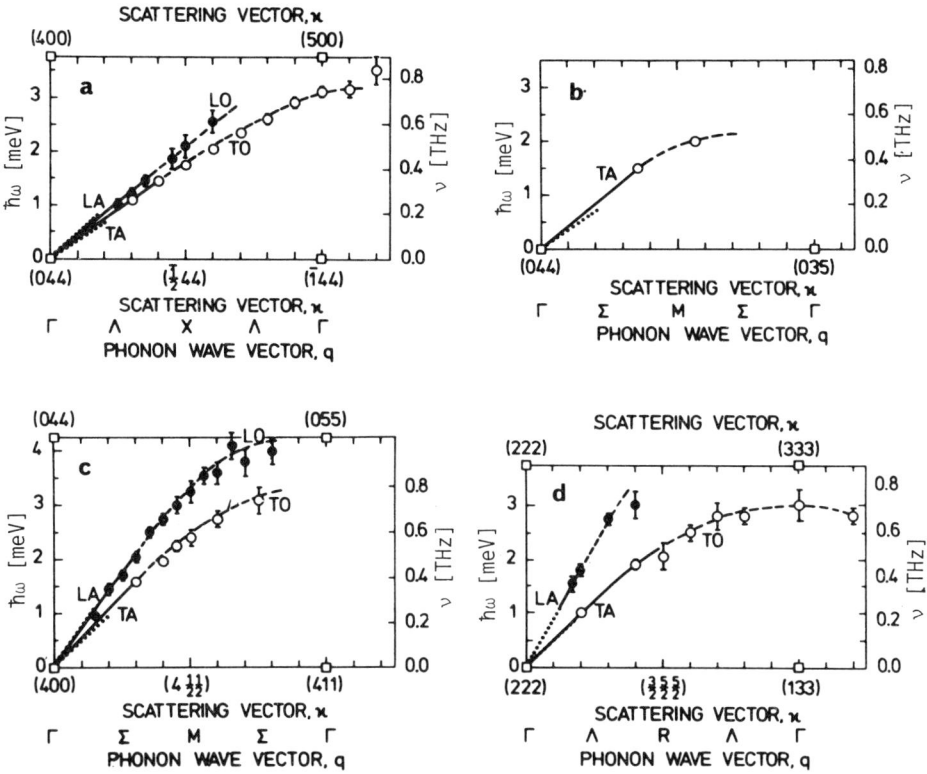

Fig. 17.8. SnI$_4$: ω(q) [Ref. 17.9, Fig. 3], T = RT, M: -

NaNO$_2$

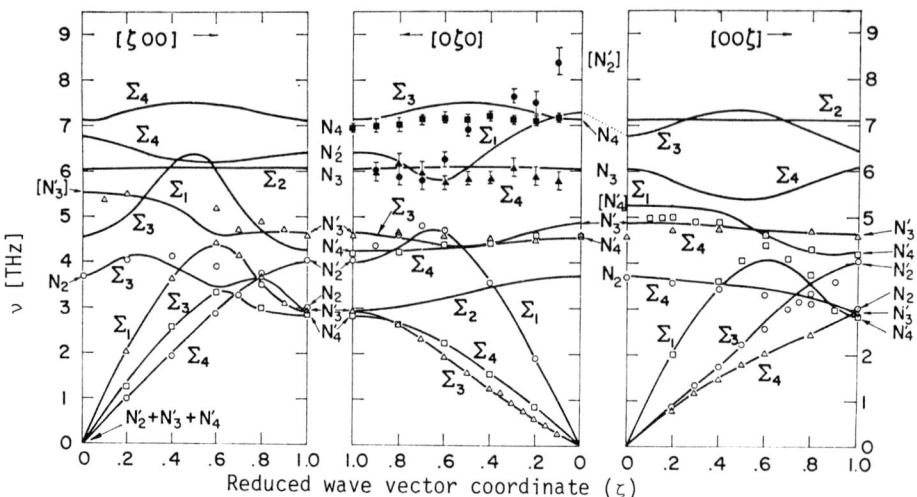

Fig. 17.9. NaNO$_2$: $\omega(q)$ [Ref.17.10, Fig.2], T = 296 K (open points), T = 110 K, (filled points), M: 22P-FCM

18. Mixed Crystals

The few measured mixed crystals are still waiting for a proper theoretical analysis. A tempting idea is the possible correlation to phonons in the perfect crystals by an interpolation scheme which considers the effect of symmetry breaking on high-symmetry phonons such as those at the L or X point in the BZ of rock salt lattice [18.5,8]. Here, the understanding of the linewidth seems of major importance.

The following systems are treated:

Crystal	Figures showing	
	Dispersion curve $\omega(q)$	Density of states $D(\omega)$
$Pb_{0.8}Sn_{0.2}Se$	18.1	
$Pb_{0.87}Sn_{0.13}Te$	18.2	
$NH_4Cl_{1-x}Br_x$	18.3	
$(KBr)_{1-x}(RbBr)_x$	18.4	
$K_{0.5}Rb_{0.5}I$	18.5	

$Pb_{0.8}Sn_{0.2}Se$

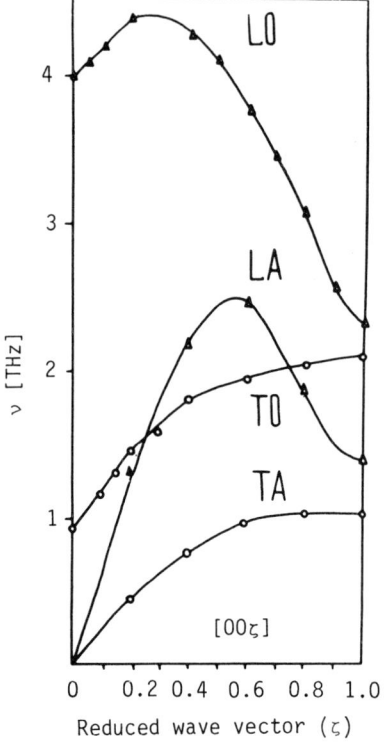

Fig. 18.1. $Pb_{0.8}Sn_{0.2}Se$: ω(q) [Ref.18.1, Fig.2], T = 80 K, M: -

Pb$_{0.87}$Sn$_{0.13}$Te

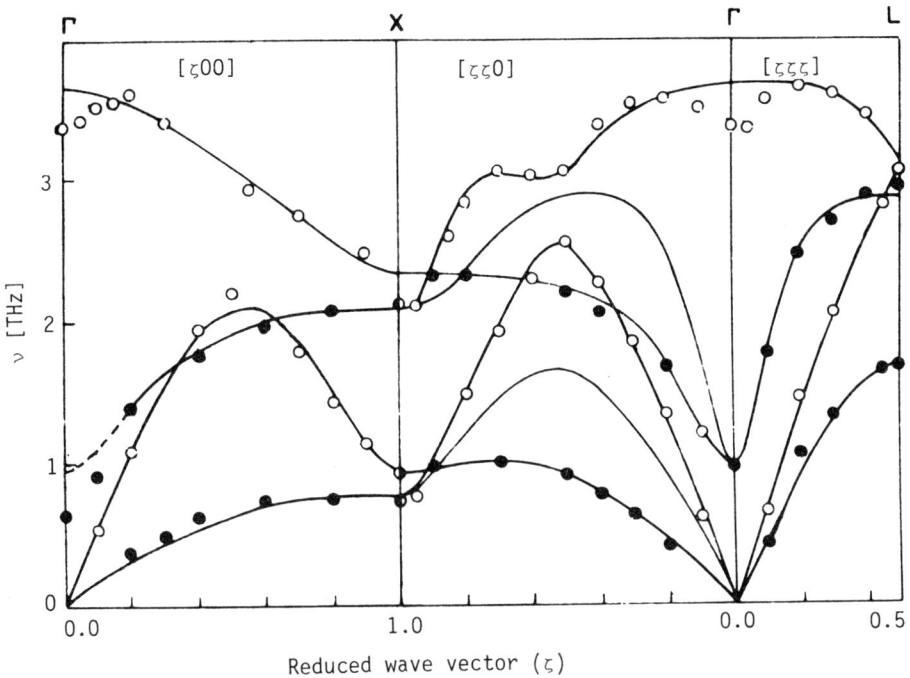

Fig. 18.2. Pb$_{0.87}$ Sn$_{0.13}$ Te: $\omega(\underline{q})$ [Ref. 18.2, Fig.1], T = 82 K, M: 15P-SM

NH Cl$_{1-x}$Br$_x$

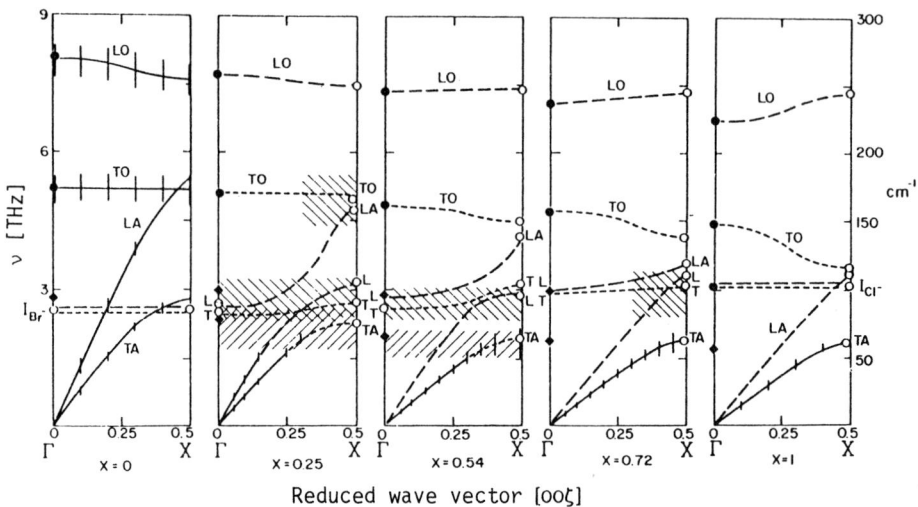

Fig. 18.3. NH$_4$Cl$_{1-x}$Br$_x$: $\omega(\underline{q})$ [Ref. 18.3, Fig.4], T = RT, | neutron peaks, \\\ neutron bands, ● IR, ◆ Raman, o---o LO [18.4], o-----o TO [18.4], —— guide lines to the eye, x = 0 [18.5], x = 1 [18.6]

$(KBr)_{rx}(RbBr)_x$

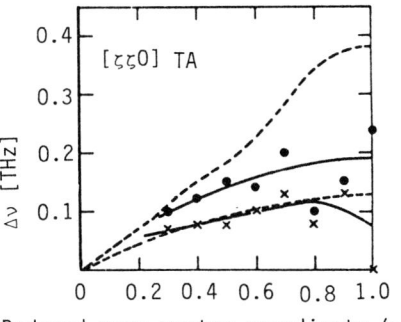
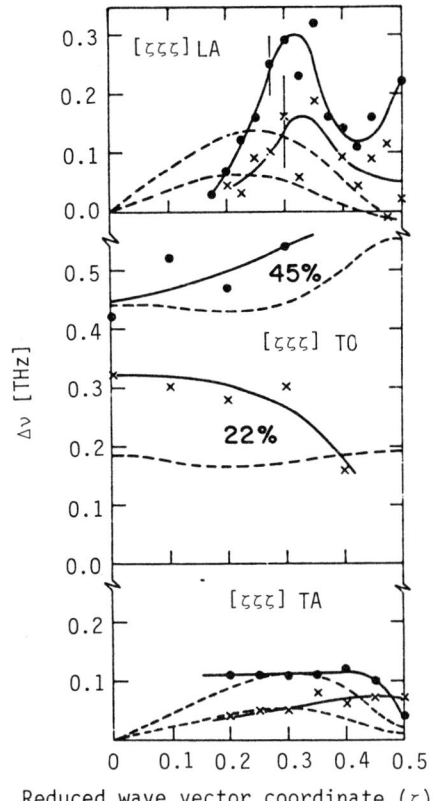

Fig. 18.4. $(KBr)_{rx}(RbBr)_x$: $\Delta\omega(\underline{q})$ [Ref. 18.7, Fig. 1], T = RT, M: 5P-SM, x = 0.22 crosses, x = 0.45 circles, guide lines to the eye (solid lines), calculations (dashed lines)

$K_{0.5}Rb_{0.5}I$

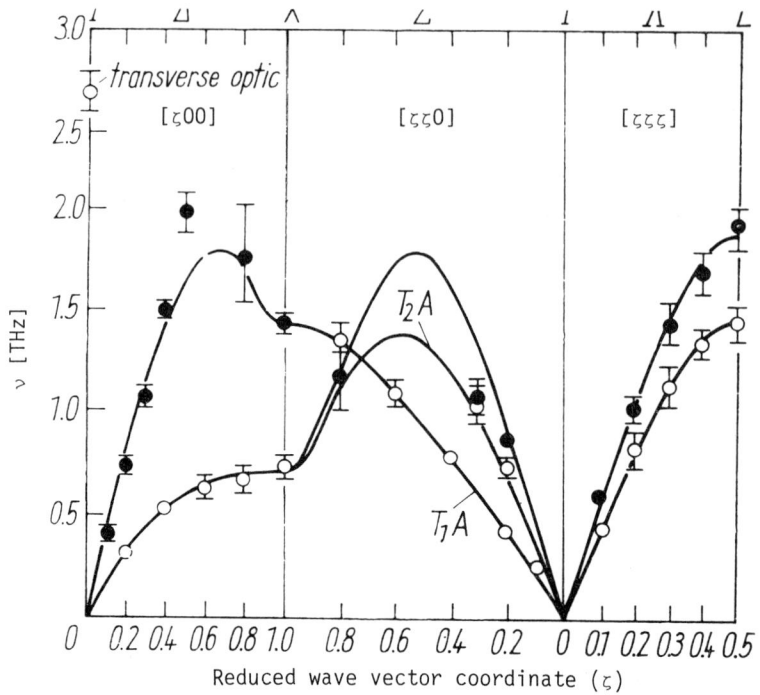

Fig. 18.5. $K_{0.5}Rb_{0.5}I$: $\omega(\underline{q})$ [Ref. 18.8, Fig. 3], T = 300 K, M: 11P-SM, Lit. [18.9]

19. Organic Crystals

In the past two decades many organic compounds have been investigated by inelastic neutron scattering. Most of these studies were time of flight measurements of the incoherent scattering from hydrogen atoms in these compounds which yield the density of states weighted by the hydrogen amplitudes. In this book incoherent neutron scattering spectra of organic compounds have not been included since it is extremely difficult to deduce the proper one-phonon density of states from these spectra.

Phonon dispersion curves have been determined by coherent inelastic neutron scattering only in a few organic compounds due to difficulties in growing suitable single crystals. The theoretical description of the measured dispersion curves uses mainly sets of formal force constants, but there is no doubt that this area will attract more sophisticated theoretical work in the near future.

The following systems are treated:

Crystal		Figures showing	
		Dispersion curve $\omega(q)$	Density of states $D(\omega)$
CD_4	(deutero-) methane	19.1a,b	
$C_4N_2D_4$	pyrazine	19.2	
$C_6D_4Cl_2$	p-dichloro-benzene	19.3	
$C_6D_{12}N_4$	hexa-methylen-tetramine	19.4a	19.4b
$C_5D_6O_2N_2$	methyl-thymine	19.5	
$C_4D_5ON_3$	cytosine (-hydrate)	19.6	
$C_{10}D_8$	naphtalene	19.7a	
$C_{14}D_{10}$	anthracene	19.8	
$CO(ND_2)_2$	urea	19.9	

CD$_4$ (Deutero-)methane

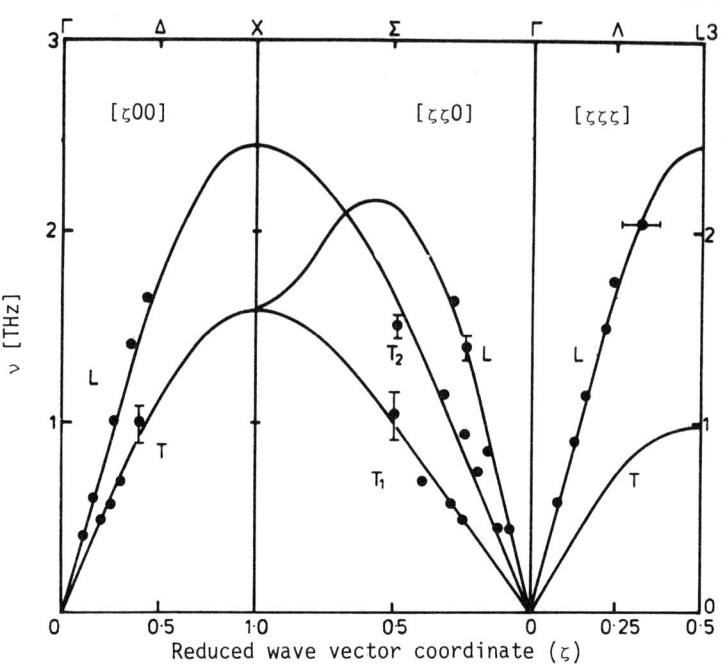

Fig. 19.1a. Methane (CD$_4$): $\omega(\underline{q})$ [Ref. 19.1, Fig. 5], T = 32,5 K (phase I), M: 3P-FCM, Lit. [19.2-5]

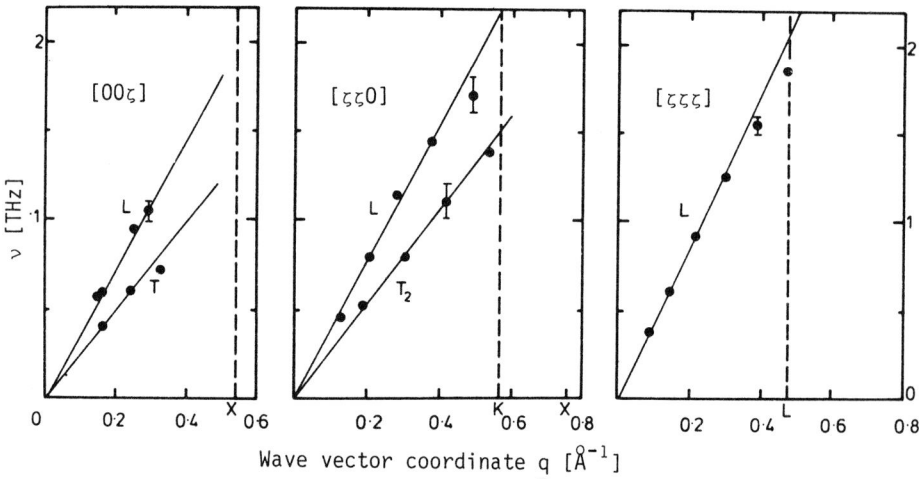

Fig. 19.1b. Methane (CD$_4$): $\omega(\underline{q})$ [Ref. 19.1, Fig. 4], T = 22.8 K (phase II)

Pyrazine $C_4N_2D_4$

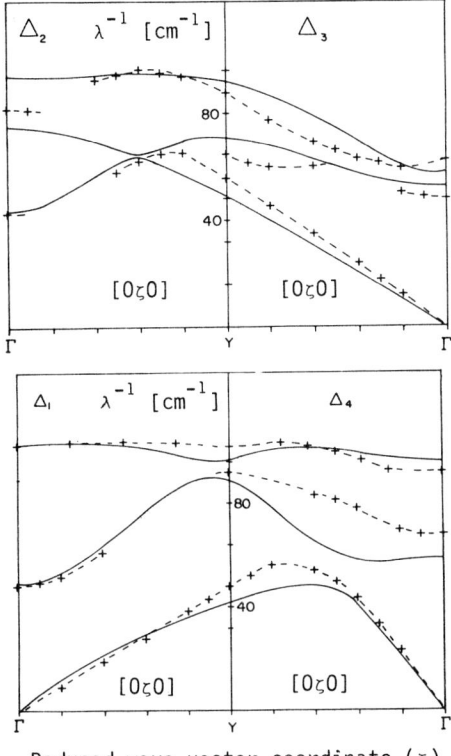

Reduced wave vector coordinate (ζ)

Fig. 19.2. Pyrazine ($C_4N_2D_4$): $\omega(\underline{q})$ [Ref. 19.6, Fig.3], T = 295 K, M: 19P-6-exp.-pot.-M (including electrical dipole moment)

$C_6D_4Cl_2$ p-Dichlorbenzene

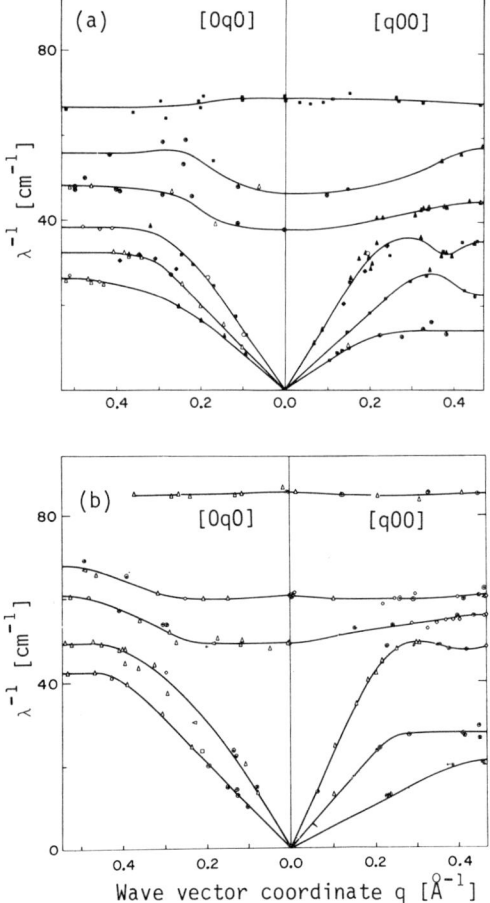

Fig. 19.3. p-dichlorobenzene ($C_6D_4Cl_2$): $\omega(\underline{q})$ [Ref. 19.7, Fig. 4], T = 295 K (a), T = 90 K (b), M: -

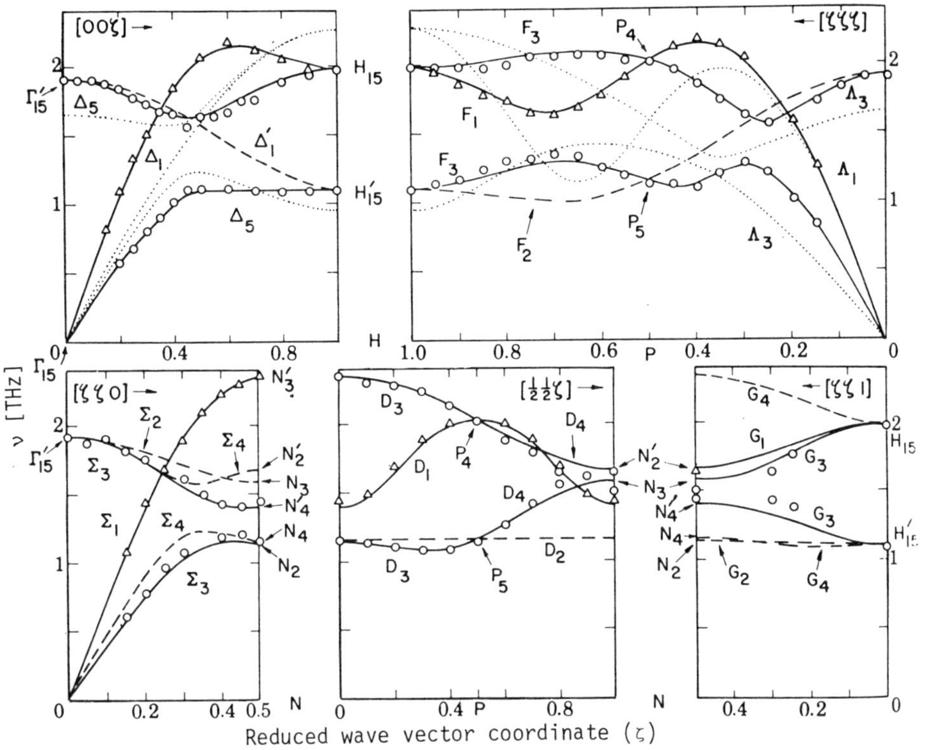

Hexamethylen-tetramine $C_6D_{12}N_4$

Fig. 19.4a. Hexamethylen-tetramine ($C_6D_{12}N_4$): $\omega(q)$ [Ref. 19.8, Fig. 2], T = 100 K, M: 5P-FCM (solid and dashed lines), 4P-FCM (dotted lines),
Lit. [19.9-12]

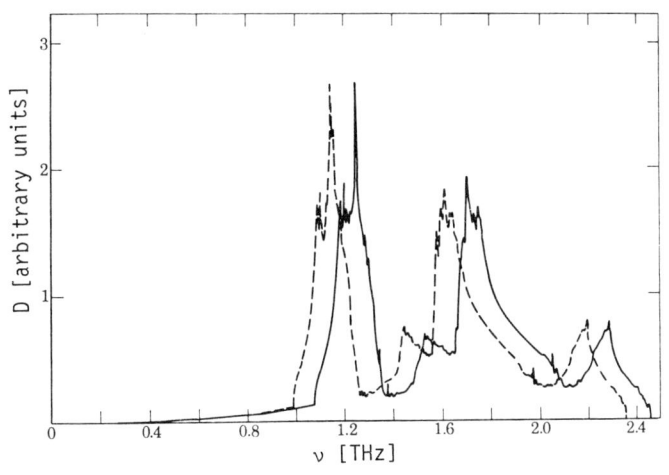

Fig. 19.4b. Hexamethylen-tetramine (HMT): $D(\omega)$ [Ref. 19.8, Fig. 5], T = 100K, M: 5P-FCM, HMT (solid lines), deuterated HMT (dashed lines)

$C_5D_6O_2N_2$ Methylthymine

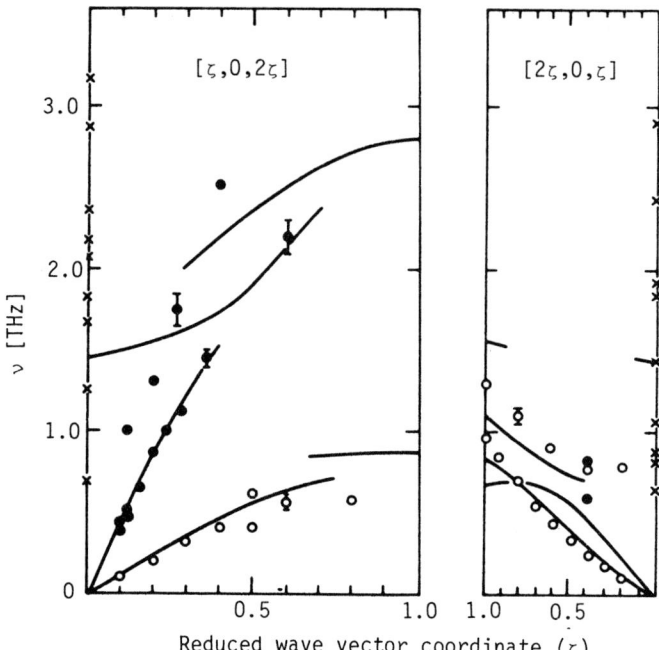

Fig. 19.5.
1-methylthymine
($C_5D_6O_2N_2$): $\omega(\underline{q})$
[Ref. 19.13,
Fig. 2], T = 296 K,
M: 28P-ASM,
Lit. [19.14]

Cytosine(-hydrate) — $C_4D_5ON_3$

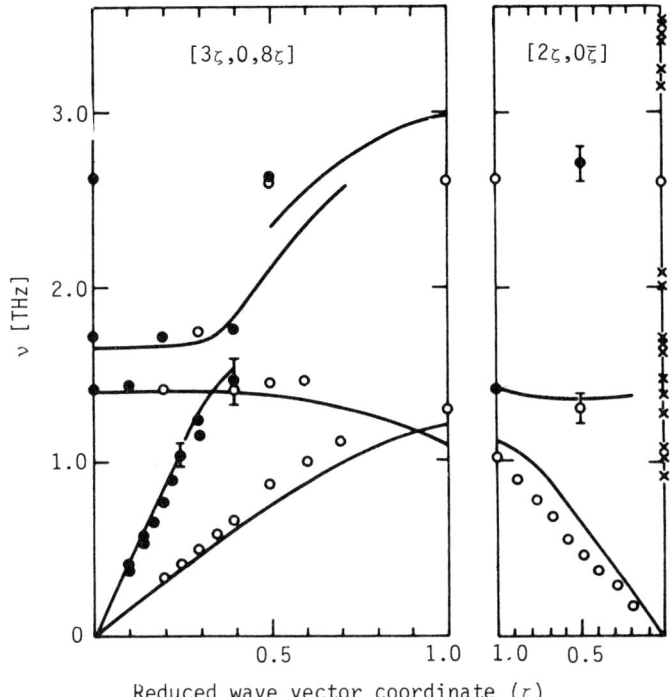

Fig. 19.6. cytosine monohydrate ($C_4D_5ON_3$): $\omega(q)$ [Ref. 19.13, Fig. 3], T = 296 K, M: 28P-ASM

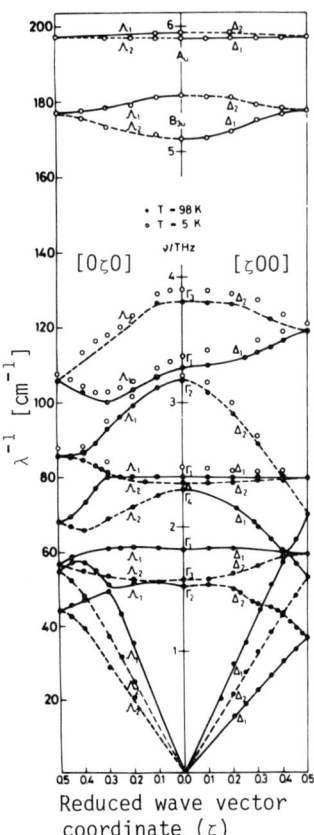

Fig. 19.7a. Naphthalene ($C_{10}D_8$): $\omega(q)$ [Ref. 19.15, Fig. 2a], T = 98 K, 5 K, M: -, Lit. [19.16-21]

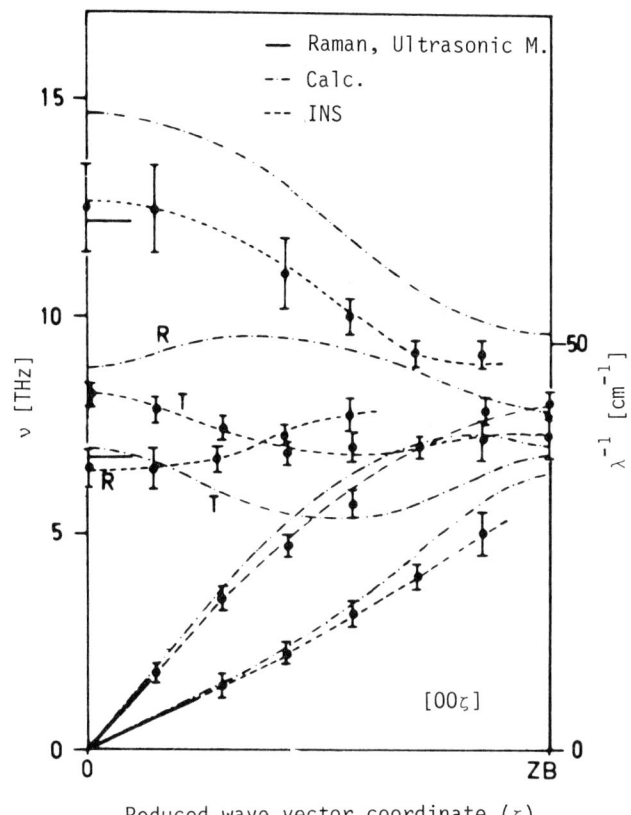

Fig. 19.8.
Anthracene ($C_{14}D_{10}$):
$\omega(\underline{q})$ [Ref. 19.22,
Fig. 1], T = RT,
M: 7P-FCM

CO(ND$_2$)$_2$ Urea

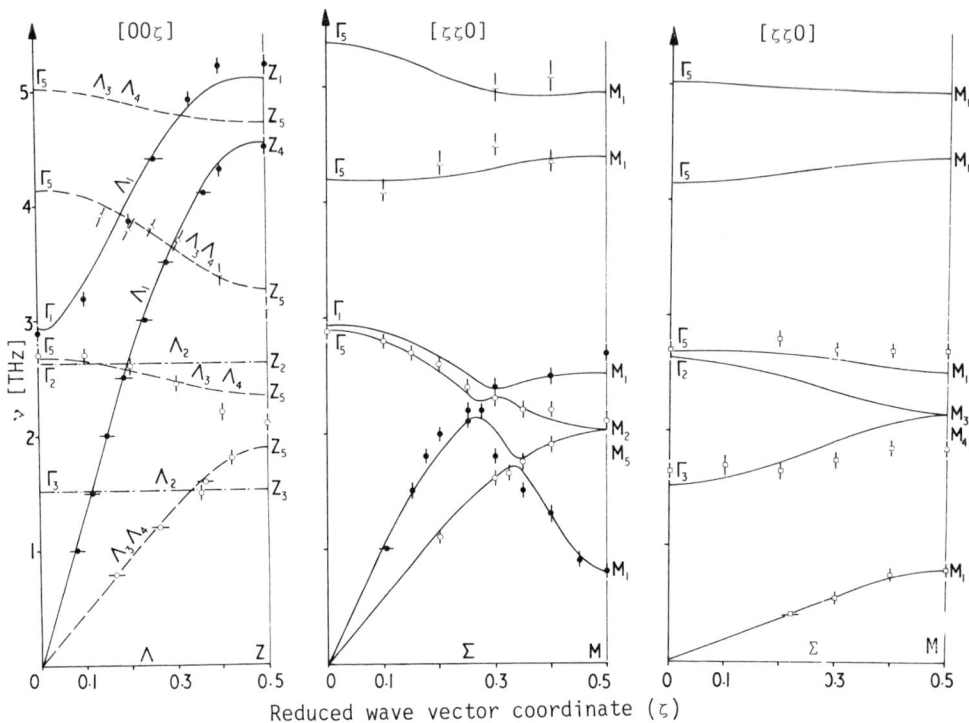

Fig. 19.9. Urea (CO(ND$_2$)$_2$): $\omega(q)$ [Ref. 19.23, Fig. 4], T = RT, M: 12P-FCM (including long range dipole-dipole interactions)

References

2.1 H. Bilz, D. Strauch, R. Wehner: *Handbuch für Physik*, Vol. 25/2d (Springer Berlin, Heidelberg, New York 1979)
2.2 J.I. Birman: *Handbuch für Physik*, Vol. 25/2b (Springer Berlin, Heidelberg, New York 1974)
2.3 M. Born, K. Huang: *Dynamical Theory of Crystal Lattices* (Oxford University Press, London 1954)
2.4 G. Leibfried, W. Ludwig: Solid State Phys. *12*, 275 (1961)
2.5 R.A. Cowley: Phil. Mag. *11*, 673 (1965)
2.6 R.S. Leigh, B. Szigeti, V.K. Tewary: Proc. Roy. Soc. A *320*, 505 (1971)
2.7 A.A. Maradudin, E.W. Montroll, G.H. Weiss, I.P. Ipatova: Solid State Phys. Suppl. *3*, 1 (1972)
2.8 W. Cochran: Crit. Rev. Solid State Sci. *2*, 1 (1971)
2.9 J. Hardy in *Dynamical Properties of Solids*, Vol. 1, ed. by G.K. Horton, A.A. Maradudin (North Holland, Amsterdam 1974) p. 157
2.10 S.K. Sinha: Crit. Rev. Solid State Sci. *3*, 273 (1973)
2.11 H. Bilz, B. Gliss, W. Hanke in *Dynamical Properties of Solids*, Vol. 1, ed. by G.K. Horton, A.A. Maradudin (North Holland, Amsterdam 1974) p. 343
2.12 W. Hanke: Adv. in Phys. *27*, 287 (1978)
2.13 R.A. Cowley, W. Cochran, B.N. Brockhouse, A.D.B. Woods: Phys. Rev. *131*, 1030 (1963)
2.14 H. Bilz, M. Buchanan, K. Fischer, R. Haberkorn: Solid State Commun. *16*, 1023 (1975)
2.15 K. Kunc, H. Bilz: Solid State Commun. *19*, 1027 (1976)
2.16 S.S. Jaswal: Phys. Rev. Lett. *35*, 1600 (1975)
2.17 R. Zeyher: Phys. Rev. Lett. *35*, 174 (1975)
2.18 U. Schröder: Solid State Commun. *4*, 347 (1966)
2.19 P. Allen: Phys. Rev. B *16*, 5139 (1977)
2.20 W. Kress, H. Bilz: to be published
2.21 W. Weber: Phys. Rev. B *8*, 5082 (1973)
2.22 A.N. Basu, D. Roy, S. Sengupta: Phys. Status Solidi (a) *23*, 11 (1974)
2.23 K. Fischer, H. Bilz, R. Haberkorn, W. Weber: Phys. Status Solidi (b) *54*, 285 (1972)
2.24 B. Dorner, W. von der Osten, W. Bührer: J. Phys. C *9*, 723 (1976)
2.25 W. Kleppmann, W. Weber: to be published
2.26 D.H. Kühner, H.V. Lauer, W.E. Bron: Phys. Rev. B *5*, 4112 (1972)
2.27 K.B. Tolpygo: Ukr. Fiz. Zh *18*, 1636 (1973); O.N. Bolonin, K.B. Tolpygo: Sov. Phys. Solid State *18*, 446 (1976)
2.28 W. Cochran: Proc. Roy. Soc. (London) A *253*, 260 (1959)
2.29 W. Kress: Phys. Status Solidi (b) *49*, 235 (1972)
2.30 W. Weber: Phys. Rev. B *15*, 4789 (1977)
2.31 P.N. Keating: Phys. Rev. *140* A, 369 (1965)
2.32 E.B. Wilson, jr., J.C. Decius, P.C. Cross: *Molecular Vibrations* (McGraw-Hill, New York 1955)
2.33 J.H. Schachtschneider, R.G. Snyder: Spectrochim. Acta *19*, 117 (1963)
2.34 R. Tubino, L. Piseri, G. Zerbi: J. Chem. Phys. *6*, 1022 (1972)
2.35 K. Kunc, M. Balkanski, M. Nusimovici: Phys. Rev. B *12*, 4346 (1975)

2.36 R.K. Singh, M.P. Verma: Phys. Status Solidi *36*, 335 (1969), *38*, 851 (1970)
2.37 R. Zeyher: Phys. Status Solidi (b) *48*, 711 (1971)
2.38 R.K. Singh, H.N. Gupta, M.K. Agrawal: Phys. Rev. B *17*, 894 (1978)
2.39 G.K. Horton, A.A. Maradudin (Eds.): *Dynamical Properties of Solids*, Vol. 1 (North Holland Amsterdam 1974)
2.40 R. Zeyher: Phys. Rev. Lett. *35*, 174 (1975)
3.1   C. Stassis, G. Kline, W. Kamitakara, S.K. Sinha: Phys. Rev. B *17*, 1130 (1978)
3.2   H. Horner: Solid State Commun. *9*, 79 (1971)
3.3   R.A. Reese, S.K. Sinha, T.O. Brun, C.R. Tilford: Phys. Rev. A *3*, 1688 (1971)
3.4   V.J. Minkiewicz, T.A. Kitchens, F.P. Lipschultz, R. Nathans, G. Shirane: Phys. Rev. *174*, 267 (1968)
3.5   V.J. Minkiewicz, T.A. Kitchens, G. Shirane, E.B. Osgood: Phys. Rev. A *8*, 1513 (1973)
3.6   H.R. Glyde: J. Low. Temp. Phys. *3*, 559 (1970)
3.7   H.R. Glyde: Can. J. Phys. *49*, 761 (1971)
3.8   J.A. Leake, W.B. Daniels, J. Skalyo, Jr., B.C. Frazer, G. Shirane: Phys. Rev. *181*, 1251 (1969)
3.9   J. Skalyo, Jr., V.J. Minkiewicz, G. Shirane, W.B. Daniels: Phys. Rev. B *6*, 4766 (1972)
3.10 J. Eckert, W.B. Daniels, J.D. Axe: Phys. Rev. B *14*, 3649 (1976)
3.11 Y. Endoh, G. Shirane, J. Skalyo, Jr.: Phys. Rev. B *11*, 1681 (1975)
3.12 Y. Fujii, N.A. Lurie, R. Pynn, G. Shirane: Phys. Rev. B *10*, 3647 (1974)
3.13 J.A. Barker, M.L. Klein, M.V. Bobetic: Phys. Rev. B *2*, 4176 (1970)
3.14 J. Skalyo, Jr., Y. Endoh, G. Shirane: Phys. Rev. B *9*, 1797 (1974)
3.15 J.A. Barker, M.V. Bobetic, M.L. Klein: Phys. Lett. A *34*, 415 (1971)
3.16 N.A. Lurie, G. Shirane, J. Skalyo, Jr.: Phys. Rev. B *9*, 5300 (1974)
3.17 B.J. Palmer, D.H. Saunderson, D.N. Batchelder: J. Phys. C *6*, L313 (1973)
4.1   J.L. Verble, J.L. Warren, J.L. Yarnell: Phys. Rev. *168*, 980 (1968)
4.2   M.P. Verma, R.K. Singh: J. Phys. C *4*, 2749 (1971)
4.3   R.K. Singh: J. Phys. C *7*, 3473 (1974)
4.4   R. Zeyher: Phys. Rev. Lett. *35*, 174 (1975)
4.5   R. Zeyher in Proc. Conf. Neutron Scattering, Vol. 1 (Gatlinburg, USA 1976) p. 550
4.6   S.S. Jaswal, V.D. Dilly: Phys. Rev. B *15*, 2366 (1977)
4.7   D. Laplace: J. Phys. C *10*, 3499 (1977)
4.8   R. Zeyher in *Lattice Dynamics*, ed. by M. Balkanski (Flammarion, Paris 1978) p.17
4.9   M.G. Zemlianov, E.G. Brovman, N.A. Chernoplekov, Iu.L. Shitikov in *Inelastic Scattering of Neutrons*, Vol. 2 (IAEA, Vienna 1965) p. 431
4.10 G. Dolling, H.G. Smith, R.M. Nicklow, P.R. Vijayaraghavan, M.K. Wilkinson: Phys. Rev. *168*, 970 (1968)
4.11 V. Nüßlein, U. Schröder: Phys. Status Solidi *21*, 309 (1967)
4.12 A.N. Basu, S. Sengupta: J. Phys. C *5*, 1158 (1972)
4.13 A. Rastogi, J.P. Hawranek, R.P. Lowndes: Phys. Rev. B *9*, 1938 (1974)
4.14 R.K. Singh: J. Phys. C *7*, 3473 (1974)
4.15 A.D.B. Woods, B.N. Brockhouse, M. Sakomoto, R.N. Sinclair: *Inelastic Scattering of Neutrons in Solids and Liquids*, Vol. 1 (IAEA, Vienna 1961) p. 487
4.16 W.J.L. Buyers: Phys. Rev. *153*, 923 (1967)
4.17 I.S. Braude: Sov. Phys. Solid State *13*, 327 (1971)
4.18 J.S. Melvin, J.D. Pirie, T. Smith: Phys. Rev. *175*, 1082 (1968)
4.19 A.M. Karo, J.R. Hardy: Phys. Rev. *181*, 1272 (1969)
4.20 R.K. Singh, M.P. Verma: Phys. Rev. B *2*, 4289 (1970)

4.21 K.V. Namjoshi, S.S. Mitra, J.F. Vetelino: Solid State Commun. *9*, 185 (1971)
4.22 A. Ghosh, A.N. Basu, S. Sengupta: Proc. Roy. Soc. (GB) A *340*, 199 (1974)
4.23 Sneh, B. Dayal: Phys. Status Solidi (b) *67*, 125 (1975)
4.24 R.K. Singh, K. Chandra: Phys. Rev. B *14*, 2625 (1976)
4.25 R.W. MacPherson, T. Timusk: Can. J. Phys. *48*, 2917 (1970)
4.26 G. Raunio, S. Rolandson: Phys. Rev. B *2*, 2098 (1970)
4.27 G. Raunio, L. Almquist, R. Stedman: Phys. Rev. *178*, 1496 (1969)
4.28 R.E. Schmunk, D.R. Winder: J. Phys. Chem. Solids *31*, 131 (1970)
4.29 E.R. Cowley: J. Phys. C *5*, 1345 (1972)
4.30 O.N. Bolonin: Sov. Phys. Solid State *18*, 1415 (1976)
4.31 O.N. Bolonin, K.B. Tolpygo: Sov. Phys. Solid State *18*, 446 (1976)
4.32 S.K. Sarkar, S. Sengupta: Phys. Status Solidi (b) *87*, 517 (1978)
4.33 A.N. Basu, S. Sengupta in *Lattice Dynamics*, ed. by M. Balkanski (Flammarion, Paris 1978) p.43
4.34 J.S. Reid, T. Smith, W.J.L. Buyers: Phys. Rev. B *1*, 1833 (1970)
4.35 O.N. Bolonin: Sov. Phys. Solid State *19*, 1088 (1977)
4.36 U. Schröder: Solid State Commun. *4*, 347 (1966)
4.37 A.D.B. Woods, B.N. Brockhouse, R.A. Cowley, W. Cochran: Phys. Rev. *131*, 1025 (1963)
4.38 A.D.B. Woods, W. Cochran, B.N. Brockhouse: Phys. Rev. *119*, 980 (1960)
4.39 R.A. Cowley, W. Cochran, B.N. Brockhouse, A.D.B. Woods: Phys. Rev. *131*, 1030 (1963)
4.40 A.N. Basu, S. Sengupta: Phys. Status Solidi *29*, 367 (1968)
4.41 T.I. Kucher, O.E. Tomasevich: Sov. Phys. Solid State *12*, 423 (1970)
4.42 I.S. Braude: Sov. Phys. Solid State *13*, 327 (1971)
4.43 A. Loidl, H. Jex, J. Daubert, M. Müllner: Phys. Status Solidi (b) *76*, 581 (1976)
4.44 D. Bäuerle, B. Fritz: Phys. Status Solidi *24*, 207 (1967)
4.45 W. Bührer: Phys. Status Solidi *41*, 789 (1970)
4.46 R.K. Singh, M.P. Verma: Phys. Status Solidi *38*, 851 (1970)
4.47 J.R.D. Copley, R.W. MacPherson, T. Timusk: Phys. Rev. *182*, 965 (1969)
4.48 G. Raunio, L. Almquist: Phys. Status Solidi *33*, 209 (1969)
4.49 N. Wakabayashi, S.K. Sinha: Phys. Rev. B *10*, 745 (1974)
4.50 A.N. Basu, S. Sengupta: Phys. Rev. B *14*, 2635 (1976)
4.51 R.A. Cowley, W. Cochran, B.N. Brockhouse, A.D.B. Woods: Phys. Rev. *131*, 1030 (1963)
4.52 E.C. Svenson, W.L.J. Buyers: Phys. Rev. *165*, 1063 (1968)
4.53 R.A. Cowley, E.C. Svenson, W.L.J. Buyers: Phys. Rev. Lett. *23*, 525 (1969)
4.54 W. Bluthardt, W. Schneider, M. Wagner: Phys. Status Solidi (b) *56*, 453 (1973)
4.55 G. Dolling, R.A. Cowley, C. Schittenhelm, J.M. Thorson: Phys. Rev. *147*, 577 (1966)
4.56 D. Bäuerle, R. Hübner: Phys. Rev. B *2*, 4252 (1970)
4.57 G. Raunio, S. Rolandson: J. Phys. C *3*, 1013 (1970)
4.58 H.H. Lal, M.P. Verma: J. Phys. C *5*, 543 (1972)
4.59 M.J.L. Sangster, R.M. Atwood: J. Phys. C *11*, 1541 (1978)
4.60 S. Rolandson, G. Raunio: J. Phys. C *4*, 958 (1971)
4.61 G. Raunio, S. Rolandson: Phys. Status Solidi *40*, 749 (1970)
4.62 W. Kress: Phys. Status Solidi (b) *62*, 403 (1974)
4.63 K.V. Namjoshi, S.S. Mitra, J.F. Vetelino in *Phonons*, ed. by M.A. Nusimovici (Flammarion, Paris 1971) p.79
4.64 K.V. Namjoshi, S.S. Mitra, J.F. Vetelino, Phys. Rev. B *3*, 4398 (1971)
4.65 R.K. Singh, H.N. Gupta, M.K. Agrawal: Phys. Rev. B *17*, 894 (1978)
5.1 M.J.L. Sangster, G. Peckham, D.H. Saunderson: J. Phys. C *3*, 1026 (1970)

5.2   K.V. Namjoshi, S.S. Mitra, J.F. Vetelino: Solid State Commun. *9*, 185 (1971)
5.3   N.S. Gillis: Phys. Rev. B *3*, 1482 (1971)
5.4   M. Caltier, A. Montaner in *Phonons*, ed. by M.A. Nusimovici, (Flammarion, Paris 1971) p.59
5.5   R.K. Singh, K.S. Upadhyaya: Phys. Rev. B *6*, 1589 (1972)
5.6   A. Ghosh, A.N. Basu in *Lattice Dynamics*, ed. by M. Balkanski (Flammarion, Paris 1978) p.65
5.7   K.H. Rieder, B.A. Weinstein, M. Cardona, H. Bilz: Phys. Rev. B *8*, 4780 (1973)
5.8   D.H. Saunderson, G. Peckham: J. Phys. C *4*, 2009 (1971)
5.9   R.P. Vijayaraghavan, Marsongkohari, P.K. Iyengar in *Neutron Inelastic Scattering* (IAEA, Vienna 1972) p.95
5.10 K.S. Upadhyaya, R.K. Singh: J. Phys. Chem. Solids *36*, 293 (1975)
5.11 K.H. Rieder, R. Migoni, B. Renker: Phys. Rev. B *12*, 3374 (1975)
5.12 S.K. Agrawal: J. Phys. Chem. Solids *38*, 199 (1977)
5.13 S.S. Chang, C.W. Tompson, E. Gürmen, L.D. Mühlestein: J. Phys. Chem. Solids *36*, 769 (1975)
5.14 W. Kress, W. Reichardt, V. Wagner, G. Kugel, B. Hennion in *Lattice Dynamics*, ed. by M. Balkanski (Flammarion, Paris 1978) p.77
5.15 B.C. Haywood, M.F. Collins: J. Phys. C *4*, 1299 (1971)
5.16 K.S. Upadhyaya, R.K. Singh: J. Phys. Chem. Solids *35*, 1175 (1974)
5.17 V. Wagner, W. Reichardt, W. Kress: *Proc. Conf. Neutron Scattering*, Vol. 1 (Gatlinburg, USA 1976) p.175
5.18 B.R.K. Gupta, M.P. Verma: J. Phys. Chem. Solids *38*, 929 (1977)
5.19 G. Kugel, C. Carabatos, B. Hennion, B. Prevot, A. Revcolevschi, D. Tocchetti: Phys. Rev. B *16*, 378 (1977)
5.20 J. Sakurai, W.J.S. Buyers, R.A. Cowley, G. Dolling: Phys. Rev. *167*, 510 (1968)
5.21 W.J.L. Buyers, G. Dolling, J. Sakurai, R.A. Cowley in *Neutron Inelastic Scattering*, Vol. 2 (IAEA Vienna 1968) p.123
5.22 W. Reichardt, V. Wagner, W. Kress: J. Phys. C *8*, 3955 (1975)
5.23 R.A. Coy, C.W. Tompson, E. Gürmen: Solid State Commun. *18*, 845 (1976)
5.24 R.P. Goyal: Phys. Status Solidi (b) *79*, K115 (1977)
5.25 M.P. Verma, S.K. Agraval: Phys. Rev. B *8*, 4880 (1973)
5.26 T.S. Chen, F.W. DeWette, L. Kleinman, D.G. Dempey: Phys. Rev. B *17*, 844 (1978)
6.1   P. Roedhammer, W. Reichardt, F. Holtzberg: Phys. Rev. Lett. *40*, 465 (1978)
6.2   P. Roedhammer, W. Reichardt, F. Holtzberg in *Lattice Dynamics*, ed. by M. Balkanski (Flammarion, Paris 1978) p.84
6.3   L. Pintschovius, W. Reichardt, B. Scheerer: J. Phys. C *11*, 1557 (1978)
6.4   W. Kress, P. Roedhammer, H. Bilz, W.D. Teuchert, A.N. Christensen: Phys. Rev. B *17*, 111 (1978)
6.5   K.H. Rieder, W. Drexel: Phys. Rev. Lett. *34*, 148 (1975)
6.6   W. Weber: Phys. Rev. B *8*, 5082 (1973)
6.7   H.G. Smith in *Superconductivity in d and f Band Transition Metals*, ed. by D.H. Douglass (AIP, New York 1972) p.321
6.8   H.G. Smith, N. Wakabayashi, R.M. Nicklow, S. Mihailovich in: *Proc. 13th Int. Conf. on Low Temperature Physics*, ed. by R.H. Kropschot, K.P. Timmerhaus (Boulder, USA 1972)
6.9   B.R.K. Gupta, M.P. Verma: J. Phys. Chem. Solids *37*, 815 (1976)
6.10 F. Gompf, J. Salgado, W. Reichardt: KFK-Report 2054 (1973/74) p.21
6.11 F. Gompf: KFK-Report 2357 (1975/76) p.9
6.12 H.G. Smith, W. Gläser: Phys. Rev. Lett. *25*, 1611 (1970)
6.13 W. Weber, H. Bilz, U. Schröder: Phys. Rev. Lett. *28*, 600 (1972)
6.14 M. Mostoller: Phys. Rev. B *5*, 1260 (1972)
6.15 M.P. Verma, B.R.K. Gupta: Phys. Rev. B *12*, 1314 (1975)

6.16 L. Pintschovius, W. Reichardt, B. Scheerer: KFK-Report 2538 (1977) p.4
6.17 H.G. Smith, W. Gläser in *Phonons*, ed. by M.A. Nusimovici (Flammarion, Paris 1971)
6.18 F. Gompf, L. Pintschovius, W. Reichardt, B. Scheerer in: *Proc. Conf. Neutron Scattering*, Vol. 1, ed. by R.M. Moon (Gatlinburg, USA 1976) CONF-760601, p.129
6.19 W. Hanke, J. Hafner, H. Bilz: Phys. Rev. Lett. *37*, 1560 (1976)
6.20 B. Splettstösser: Z. Phys. B *26*, 151 (1977)
6.21 A.K. Goyal, S.C. Goyal: Phys. Status Solidi (b) *81*, K63 (1977)
6.22 F. Gompf, W. Reichardt, A.N. Christensen: KFK-Report 2357 (1975/76) p.11
6.23 S.C. Goyal, A.K. Goyal: Phys. Status Solidi (b) *77*, 639 (1976)
6.24 G. Dolling, T.M. Holden, E.C. Svensson, W.J.L. Buyers, G.H. Lander in *Lattice Dynamics*, ed. by M. Balkanski (Flammarion, Paris 1978) p.81
6.25 V. Varma, W. Weber: Phys. Rev. Lett. *39*, 1094 (1977)
6.26 F. Gompf, W. Reichardt: KFK-Report 2538 (1977) p.6
7.1 E.R. Cowley, J.K. Darby, G.S. Pawley: J. Phys. C *2*, 1916 (1969)
7.2 G.S. Pawley, W. Cochran, R.A. Cowley, G. Dolling: Phys. Rev. Lett. *17*, 753 (1966)
7.3 M.M. Elcombe: Proc. Roy. Soc. (London) A *300*, 210 (1967)
7.4 P.R. Vijayaraghavan, S.K. Sinha, P.K. Iyengar in *Proc. Nuc. Phys. and Solid State Symp.* (Bangalore, India 1973) p.208
7.5 W. Cochran, R.A. Cowley, G. Dolling, M.M. Elcombe: Proc. Roy. Soc. (London) A *293*, 433 (1966)
7.6 H.A. Alperin, S.J. Pickart, J.J. Rhyne, V.J. Minkiewicz: Phys. Lett. A *40*, 295 (1972)
7.7 N. Wakabayashi, A. Furrer: Phys. Rev. B *13*, 4343 (1976)
7.8 P.R. Vijayaraghavan, R.M. Nicklow, H.G. Smith, M.K. Wilkinson: Phys. Rev. B *1*, 4819 (1970)
7.9 R.K. Singh, K.S. Upadhyaya: Phys. Status Solidi (b) *51*, 389 (1972)
7.10 K. Fischer, H. Bilz, R. Haberkorn, W. Weber: Phys. Status Solidi (b) *54*, 285 (1972)
7.11 K.S. Upadhyaya, P.S. Mahesh: Phys. Status Solidi (b) *59*, 279 (1973)
7.12 J.P. Hawranek, R.P. Lowndes: Solid State Commun. *11*, 1473 (1972)
7.13 H. Kanzaki, S. Sakurayi, S. Hoshino, J. Shirane, Y. Fujii: Solid State Commun. *15*, 1547 (1974)
7.14 B. Dorner, W. von der Osten, W. Bührer: J. Phys. C *9*, 723 (1976)
7.15 W. von der Osten, B. Dorner: Solid State Commun. *16*, 431 (1975)
7.16 W. Bührer: Phys. Status Solidi (b) *68*, 739 (1975)
7.17 R.K. Singh, K.P. Singh: Phys. Status Solidi (b) *78*, 677 (1976)
7.18 Y. Fujii, S. Hoshino, S. Sakurayi, H. Kanzaki, J.W. Lynn, G. Shirane: Phys. Rev. B *15*, 358 (1977)
7.19 B. Dorner, J. Windscheif, W. von der Osten in: *Lattice Dynamics*, ed. by M. Balkanski, (Flammarion, Paris 1978) p.535
7.20 R. Zeyher in: *Lattice Dynamics*, ed. by M. Balkanski (Flammarion, Paris 1978) p.17
7.21 J.M. Rowe, J.J. Rush, N. Vagelatos, D.L. Price, D.G. Hinks, S. Susman: J. Chem. Phys. *62*, 4551 (1975)
7.22 C.J. Bill, H. Jex, M. Müllner: Phys. Lett. A *56*, 320 (1976)
7.23 H. Jex, C.J. Maetz: Phys. Status Solidi (b) *85*, 511 (1978)
7.24 J. Daubert, K. Knorr, W. Dultz, H. Jex, R. Currat: J. Phys. C *9*, L389 (1976)
7.25 D. Fontaine, H. Poulet: Phys. Status Solidi (b) *58* K9 (1973)
7.26 W. Dultz: Solid State Commun. *15*, 595 (1974)
7.27 N. Nücker, K. Knorr, H. Jex: J. Phys. C *11*, 1 (1977)
7.28 N. Vagelatos, J.M. Rowe, J.J. Rush: Phys. Rev. B *12*, 4522 (1975)

7.29 C.M. Goel. T.P. Sharma, B. Dayal: J. Phys. Chem. Solids *38*, 1285 (1977)
7.30 S. Rolandson: Phys. Status Solids (b) *52*, 643 (1972)
7.31 W. Bühner: J. Phys. C *6*, 2931 (1973)
7.32 H.N. Gupta, R.K. Singh: Phys. Status Solidi (b) *61*, 681 (1974)
7.33 M.J.L. Sangster, R.M. Atwood: J. Phys. C *11*, 1541 (1978)
7.34 J.M. Rowe, J.J. Rush, E. Prince, N.J. Chesser: Ferroelectrics *16*, 107 (1977)
7.35 P.R. Vijayaraghavan, S.K. Sinha, P.K. Iyengar in: *Annual Report of the Nuclear Physics Division*, Bhabha Atomic Research Center, Bombay (1976) p.69
7.36 P.R. Vijayaraghavan: private communication (1978)
7.37 R.J. Birgeneau, S.M. Shapiro in: *Valence Instabilities and Narrow-Band Phenomena*, ed. by R.D. Parks (Plenum Press, New York 1977) p.49
7.38 H.A. Mook, R.M. Nicklow, T. Penney, F. Holtzberg, W.M. Shafer: Phys. Rev. B *18*, 2925 (1978)
8.1 A.A. Ahmad, H.G. Smith, N. Wakabayashi, M.K. Wilkinson: Phys. Rev. B *6*, 3956 (1972)
8.2 J.F. Vetelino, K.V. Namjoshi, S.S. Mitra: Phys. Rev. B *7*, 4001 (1973)
8.3 A. Ghosh, A.N. Basu: J. Phys. C *9*, 4365 (1976)
8.4 M.J.L. Sangster, R.M. Atwood: J. Phys. C *11*, 1541 (1978)
8.5 S. Rolandson, G. Raunio: Phys. Rev. B *4*, 4617 (1971)
8.6 G. Mahler, P. Engelhardt: Phys. Status Solidi (b) *45*, 543 (1971)
8.7 J. Daubert in: *Neutron Inelastic Scattering* (IAEA Vienna 1972) p.85
8.8 H.H. Lal, M.P. Verma: J. Phys. C *5*, 1038 (1972)
8.9 J. Daubert, H. Jex, M. Müllner: Phys. Status Solidi (b) *57*, 477 (1973)
8.10 W. Bührer, W. Hälg: Phys. Status Solidi (b) *46*, 679 (1971)
8.11 J.S. Eldridge: J. Phys. C *3*, 1527 (1970)
8.12 B.I. Gorbachev, P.G. Ivanitskii, V.T. Krotenko, M.V. Pasechnik: Ukr. Fiz. Zh. (USSR) *18*, 92 (1973)
8.13 H.G. Smith in: G. Venkataraman, V.C. Sahni: Rev. Mod. Phys. *42*, 409 (1970)
8.14 H.Jex: Solid State Commun. *9*, 2057 (1971)
8.15 H.G. Smith, J.G. Traylor, W. Reichardt: Phys. Rev. *181*, 1218 (1969)
8.16 E.R. Cowley: Phys. Rev. B *3*, 2743 (1971)
8.17 C.H. Kim, H.A. Rafizadeh, S. Yip: J. Chem. Phys. *57*, 2291 (1972)
8.18 H.C. Teh: Can. J. Phys. *50*, 2807 (1972)
8.19 H.C. Teh.,B.N. Brockhouse: Phys. Rev. B *3*, 2733 (1971)
8.20 H. Jex: Phys. Lett. A *34*, 118 (1971)
8.21 H.C. Teh, B.N. Brockhouse: Phys. Rev. B *8*, 3928 (1973)
8.22 R.K. Singh, H.N. Gupta, M.K. Agrawal: Phys. Rev. B *17*, 894 (1978)
8.23 Y. Yamada, Y. Noda, J.D. Axe, G. Shirane: Phys. Rev. B *9*, 4429 (1974)
8.24 Y Fujii, T. Sakuma, J. Nakahara, S. Hoshino, K. Kobayashi, A. Fujii: J. Phys. Soc. Jpn. *44*, 1237 (1978)
8.25 E.R. Cowley, A. Okazaki: Proc. Roy. Soc. (London) A *300*, 45 (1967)
8.26 R. Srinivasan, G. Lakshmi, V. Ramachandran: J. Phys. C *8*, 2889 (1975)
9.1 J.L. Warren, J.L. Yarnell, G. Dolling, R.A. Cowley: Phys. Rev. *158*, 805 (1967)
9.2 R. Wehner, H. Borik, W. Kress, A. Goodwin, S.D. Smith: Solid State Commun. *5*, 307 (1967)
9.3 G. Dolling, R.A. Cowley: Proc. Phys. Soc. (London) *88*, 463 (1966)
9.4 G. Peckham: Solid State Commun. *5*, 311 (1967)
9.5 H.L. Mc Murry, A.W. Solbrig, J.K. Boyter, C. Noble: J. Phys. Chem. Solids *28*, 2359 (1967)
9.6 R. Blanchard, Y.P. Varshni: Phys. Rev. *159*, 599 (1967)
9.7 W. Cochran: Crit. Rev. Solid State Sci. *2*, 1 (1971)
9.8 S.K. Sinha: Crit. Rev. Solid State Sci. *3*, 273 (1973)

9.9   B.P. Pandey, B. Dayal: J. Phys. C 6, 2943 (1973)
9.10 R. Tubino, L. Piseri, G. Zebri: J. Chem. Phys. 56, 1022 (1972)
9.11 F.A. Johnson, K. Moore: Proc. Soc. (London) A 339, 85 (1974)
9.12 W. Weber: Phys. Rev. B 15, 4789 (1977)
9.13 G. Dolling: *Inelastic Neutron Scattering*, Vol. 2 (IAEA Vienna 1965) p.37
9.14 G. Nilsson, G. Nelin: Phys. Rev. B 6, 3777 (1972)
9.15 B.N. Brockhouse: Phys. Rev. Lett. 2, 256 (1959)
9.16 R.M. Martin: Phys. Rev. 186, 871 (1969)
9.17 W. Solbrig, Jr.: Phys. Chem. Solids 32, 1761 (1971)
9.18 D.L. Price, S.K. Sinha, R.P. Gupta: Phys. Rev. B 9, 2573 (1974)
9.19 T. Soma: Phys. Status Solidi (b) 76, 753 (1976)
9.20 G. Nilsson, G. Nelin: Phys. Rev. B 3, 364 (1971)
9.21 G. Nelin, G. Nilsson: Phys. Rev. B 5, 3151 (1972)
9.22 N.K. Pope in: *Lattice Dynamics*, ed. by R.F. Wallis (Pergamon Press, Oxford 1965) p.147
9.23 B.N. Brockhouse, P.K. Ivengar: Phys. Rev. 111, 747 (1958)
9.24 F. Herman: J. Phys. Chem. Solids 8, 405 (1959)
9.25 W. Cochran: Proc. Roy. Soc. (London) A 253, 260 (1959)
9.26 W. Kress: Phys. Status Solidi (b) 49, 235 (1972)
9.27 B.P. Pandy, B. Dayal: Solid State Commun. 11, 775 (1972)
9.28 G. Nelin: Phys. Rev. B 10, 4331 (1974)
9.29 W. Weber: Phys. Rev. Lett. 33, 371 (1974)
9.30 A.M. Al'tschuler, Yu.Kh. Vekilo, T.I. Kacherets: Sov. Phys. Solid State 18, 1279 (1976)
9.31 S.S. Jaswal in: *Lattice Dynamics*, ed. by M. Balkanski (Flammarion, Paris 1978) p.41
9.32 D.L. Price, J.M. Rowe, R.N. Nicklow: Phys. Rev. B 3, 1268 (1971)
9.33 D.L. Price, J.M. Rowe: Solid State Commun. 7, 1433 (1969)
9.34 R.P. Gupta, S.K. Sinha, J.P. Walter, M.L. Cohen: Solid State Commun. 14, 1313 (1974)
9.35 T. Soma: J. Phys. Soc. Jpn. 36, 1301 (1974)
10.1 K. Kunc, M. Balkanski, M.A. Nusimovici: Phys. Status Solidi (b) 72, 229 (1975)
10.2 D.W. Feldman, J.H. Parker, W.J. Choyke, L. Patrick: Phys. Rev. 173, 787 (1968)
10.3 J.F. Vetelino, S.S. Mitra: Phys. Rev. 178, 1349 (1969)
10.4 R. Banerjee, Y.P. Varshni: J. Phys. Soc. Jpn. 30, 1015 (1971)
10.5 T.N. Singh, S.S. Kushwaha, G. Singh: Solid State Commun. 13, 1393 (1973)
10.6 T.N. Singh, B.N. Roy: Nuovo Cimento B 41, 198 (1977)
10.7 R. Banerjee, Y.P. Varshni: Can. J. Phys. 47, 451 (1969)
10.8 L.L. Yarnell, J.L. Warren, R.G. Wenzel, P.J. Dean: *Neutron Inelastic Scattering*, Vol. 1, (IAEA, Vienna 1968) p.301
10.9 R. Banerjee, Y.P. Varshni: J. Phys. Soc. Jpn. 30, 1015 (1971)
10.10 K. Kunc, M. Balkanski, M.A. Nusimovici: Phys. Status Solidi (b) 72, 229 (1975)
10.11 K. Kunc, H. Bilz in: *Proc. Conf. Neutron Scattering*, Vol. 1, ed. by R.M. Moon (Gatlinburg, USA 1976), CONF-760601-P1 and Solid State Commun. 19, 1027 (1976)
10.12 K.C. Rustagi, W. Weber: Solid State Commun. 18, 1027 (1976)
10.13 J.L.T. Waugh, G. Dolling: Phys. Rev. 132, 2410 (1963)
10.14 G. Dolling, R.A. Cowley: Proc. Phys. Soc. (London) 88, 463 (1966)
10.15 E.N. Korol: Sov. Phys. Solid State 12, 497 (1970)
10.16 M.K. Farr, J.G. Traylor, S.K. Sinha: Phys. Rev. B 11, 1587 (1975)
10.17 M. Vandevyver, P. Plumelle: Phys. Rev. B 17, 675 (1978)
10.18 P.H. Borcherds, G.F. Alfrey, D.H. Saunderson, A.D.B. Woods: J. Phys. C 8, 2022 (1975)

10.19 M. Vandevyver, P. Plumelle: J. Phys. Chem. Solids *38*, 765 (1977)
10.20 P.H. Borcherds, K. Kunc: J. Phys. C *11*, 4145 (1978)
10.21 D.L. Price, J.M. Rowe, R.M. Nicklow: Phys. Rev. B *3*, 1268 (1971)
10.22 R. Banerjee, Y.P. Varshni: Can. J. Phys. *47*, 451 (1969)
10.23 D.N. Talwar, B.K. Agrawal: Solid State Commun. *11*, 1691 (1972)
10.24 E.S. Koteles, W.R. Datars, G. Dolling: Phys. Rev. B *9*, 572 (1974)
10.25 K. Kunc, M. Balkanski, M.A. Nusimovici: Phys. Rev. B *12*, 4346 (1975)
10.26 S.S. Jaswal in: *Lattice Dynamics*, ed. by M. Balkanski (Flammarion, Paris 1978) p.41
10.27 J. Bergsma: Phys. Rev. Lett. A *32*, 324 (1970)
10.28 L.A. Feldkamp, D.K. Steinman, N. Vagelatos, J.S. King, G. Venkataraman: J. Phys. Chem. Solids *32*, 1573 (1971)
10.29 J.F. Vetelino, S.S. Mitra, O. Brafman, T.C. Damen: Solid State Commun. *7*, 1809 (1969)
10.30 D.N. Talwar, B.K. Agrawal: Phys. Status Solidi (b) *63*, 441 (1974)
10.31 D.N. Talwar, B.K. Agrawal: Phys. Status Solidi (b) *64*, 71 (1974)
10.32 N. Vagelatos, D. Wehe, J.S. King: J. Chem. Phys. *60*, 3613 (1974)
10.33 B. Hennion, F. Moussa, G. Pepy, K. Kunc: Phys. Lett. A *36*, 377 (1971)
10.34 P. Plumelle, M. Vandevyver: Phys. Status Solidi (b) *73*, 271 (1976)
10.35 J.M. Rowe, R.M. Nicklow, D.L. Price, K. Zanio: Phys. Rev. B *10*, 671 (1974)
10.36 V.T. Bublik, S.S. Gorelik, I.S. Smirnow: Sov. Phys. Crystallogr. *17*, 485 (1972)
10.37 B. Prevot, B. Hennion, B. Dorner: J. Phys. C *10*, 3999 (1977)
10.38 C. Carabatos, B. Hennion, K. Kunc, F. Moussa, C. Schwab: Phys. Rev. Lett. *26*, 770 (1971)
10.39 Z. Vardeny, G. Gilat in: *Lattice Dynamics*, ed. by M. Balkanski (Flammarion, Paris 1978) p.55
10.40 S. Hoshino, Y. Fujii, J. Harada, J.D. Axe: J. Phys. Soc. Jpn. *41*, 965 (1976)
10.41 B. Prevot, C. Carabatos, C. Schwab, B. Hennion, F. Moussa: Solid State Commun. *13*, 1725 (1973)
10.42 B. Hennion, F. Moussa, B. Prevot, C. Carabatos, C. Schwab: Phys. Rev. Lett. *28*, 964 (1972)
11.1 G.L. Ostheller, R.E. Schmunk, R.M. Brugger, R.J. Kearnly: *Inelastic Neutron Scattering*, Vol. 1 (IAEA Vienna 1968) p.315
11.2 R.M. Brugger, K.A. Strong, J.M. Carpenter: J. Phys. Chem. Solids *28*, 249 (1967)
11.3 R. Ramani, K.K. Mani, R.P. Singh: Phys. Rev. B 14, 2659 (1976)
11.4 R.N. Sinclair: IAEA Symp. (Chalk River 1962), Vol. 2, p.199
11.5 A.W. Hewat: Solid State Commun. *8*, 187 (1970)
11.6 K. Thoma, B. Dorner, G. Duesing, W. Wegener: Solid State Commun. *15*, 1111 (1974)
11.7 W. Wegener, S. Hautecler: Phys. Lett. A *31*, 2 (1970)
11.8 K. Kunc, H. Bilz in *Proc. Conf. Neutron Scattering*, Vol. 1, ed. by R.M. Moon (Gatlinburg, USA 1976) CONF-760601-P1, p.195 and Solid State Commun. *19*, 1027 (1976)
11.9 M.A. Nusimovici, M. Balkanski, J.L. Birman: Phys. Rev. B *1*, 595 (1970)
11.10 W. Bührer, R.M. Nicklow, P. Bruesch: Phys. Rev. B *17*, 3362 (1978)
12.1 M.M. Elcombe, A.W. Pryor: J. Phys. C *3*, 492 (1970)
12.2 M. Elcombe: J. Phys. C *5*, 2702 (1972)
12.3 A. Sadoc, F. Moussa, G. Pepy: J. Phys. Chem. Solids *37*, 197 (1976)
12.4 J.P. Hurrell, V.J. Minkiewicz: Solid State Commun. *8*, 463 (1970)
12.5 M.H. Dickens, M.T. Hutchings: J. Phys. C *11*, 461 (1978)
12.6 R.J. Kearney, T.G. Worlton, R.E. Schmunk: J. Phys. Chem. Solids *31*, 1085 (1970)

12.7 N. Wakabayashi, A.A.Z. Ahmad, H.R. Shanks, G.C. Danielson: Phys. Rev. B 5, 2103 (1972)
12.8 G. Dolling, R.A. Cowley, A.D.B. Woods: Can. J. Phys. 43, 1397 (1965)
12.9 R.A. Cowley, G. Dolling: Phys. Rev. 167, 464 (1968)
13.1 J.G. Traylor, H.G. Smith, R.M. Nicklow, M.K. Wilkinson: Phys. Rev. B 3, 3457 (1971)
13.2 R.S. Katiyar in: *Phonons*, Proc. Int. Conf. Phonons, Rennes, ed. by M.A. Nusimovici (Flammarion, Paris 1971) p.84
13.3 R. Almairac, C. Benoit: J. Phys. C 7, 2614 (1974)
13.4 G.C. Cran, M.J.L. Sangster: J. Phys. C 7, 1937 (1974)
13.5 R.S. Katiyar: J. Phys. C 3, 1693 (1970)
13.6 B.D. Rainford, J.G. Houmann, H.J. Guggenheim: *Neutron Inelastic Scattering* (IAEA Vienna 1972) p.655
13.7 P. Martel, R.A. Cowley, R.W.H. Stevenson: Can. J. Phys. 46, 1355 (1968)
13.8 R.S. Katiyar, P. Dawson, M.M. Hargreave, G.R. Wilkinson: J. Phys. C 4, 2421 (1971)
13.9 R. Almairac, J.L. Sauvajol, C. Benoit, A.M. Bon: J. Phys. C 11, 3157 (1978)
14.1 M.R. Chowdhury, G.E. Peckham, R.T. Ross, D.H. Saunderson: J. Phys. C 7, L99 (1974)
14.2 C.H. Perry, H. Buhay, A.M. Quittet, R. Currat in: *Lattice Dynamics*, ed. by M. Balkanski (Flammarion, Paris 1978) p.677
14.3 R. Currat, R. Comes, B. Dorner, E. Wiesendanger: J. Phys. C 7. 2521 (1974)
14.4 R. Comes, G. Shirane: Phys. Rev. B 5, 1886 (1972)
14.5 J.D. Axe, J. Harada, G. Shirane: Phys. Rev. B 1, 1227 (1970)
14.6 G. Shirane: Rev. Mod. Phys. 46, 437 (1974)
14.7 R. Migoni, H. Bilz, D. Bäuerle: Phys. Rev. Lett. 37, 1155 (1976)
14.8 R. Migoni, H. Bilz in: *Lattice Dynamics*, ed. by M. Balkanski (Flammarion, Paris 1978) p.650
14.9 K. Gesi, J.D. Axe, G. Shirane, A. Linz: Phys. Rev. B 5, 1933 (1972)
14.10 V.J. Minkiewicz, G. Shirane: J. Phys. Soc. Jpn. 26, 674 (1969)
14.11 W.G. Stirling: J. Phys. C 5, 2711 (1972)
14.12 R.A. Cowley: Phys. Rev. 134, A 981 (1964)
14.13 Y. Yamada, G. Shirane: J. Phys. Soc. Jpn. 26, 396 (1969)
14.14 G. Shirane, Y. Yamada: Phys. Rev. 177, 858 (1969)
14.15 R.A. Cowley, W.J.L. Buyers, G. Dolling: Solid State Commun. 7, 181 (1969)
14.16 W.G. Stirling, R. Currat: J. Phys. C 9, L519 (1976)
14.17 R. Migoni, K.H. Rieder, K. Fischer, H. Bilz: Ferroelectrics 13, 377 (1976)
14.18 R.A. Cowley in: *Lattice Dynamics*, ed. by M. Balkanski (Flammarion, Paris 1978) p.625
14.19 A.J. Maeland: J. Chem. Phys. 51, 2915 (1969)
14.20 G. Shirane, B.C. Frazer, V.J. Minkiewicz, J.A. Leake: Phys. Rev. Lett. 5, 234 (1967)
14.21 Y. Yamada, G. Shirane, A. Linz: Phys. Rev. 177, 848 (1969)
14.22 J. Harada, J.D. Axe, G. Shirane: Acta Crystallogr. A 26, 608 (1970)
14.23 J. Harada, J.D. Axe, G. Shirane: Phys. Rev. B 4, 155 (1971)
14.24 G. Shirane, J.D. Axe, J. Harada: Phys. Rev. B 2, 3651 (1970)
14.25 J.D. Axe, G. Shirane, K.A. Müller: Phys. Rev. 183, 820 (1969)
14.26 G. Shirane, J.D. Axe, J. Harada, J.P. Remeika: Phys. Rev. B 2, 155 (1970)
14.27 B. Dorner, M. Steiner: J. Phys. C 9, 15 (1976)
14.28 T.M. Holden, W.J.L. Buyers, E.C. Svensson, R.A. Cowley, M.T. Hutchings, D. Hunkin, R.W.H. Stevenson: J. Phys. C 4, 2127 (1971)

14.29 E.R. Cowley, A.K. Pant: Phys. Rev. B *8*, 4795 (1973)
14.30 K.W. Logan, S.F. Trevino, R.C. Casella, W.M. Shaw, L.D. Mühlenstein, R.D. Mical in: *Phonons* (Proc. Int. Conf., Rennes), ed. by M.A. Nusimovici (Flammarion, Paris 1971) p.104
14.31 K.R. Rao, S.L. Chaplot, P.K. Iyengar, A.H. Venkatesch, P.R. Vijayaraghavan in: *Lattice Dynamics*, ed. by M. Balkanski (Flammarion, Paris 1978) p.73
14.32 K.R. Rao, S.L. Chaplot, P.K. Iyengar, A.H. Venkatesh, P.R. Vijayaraghavan: Pramana (India) *11*, 251 (1978)
15.1 R.M. Nicklow, N. Wakabayashi, H.G. Smith: Phys. Rev. B *5*, 4951 (1972)
15.2 A. Yoshimori, Y. Kitano: J. Phys. Soc. Jpn. *11*, 352 (1956)
15.3 G. Dolling, B.N. Brockhouse: Phys. Rev. *128*, 1120 (1962)
15.4 J.A. Young, J.V. Koppel: J. Chem Phys. *42*, 357 (1965)
15.5 D.I. Page in: *Neutron Inelastic Scattering*, Vol. 1 (IAEA Vienna 1968) p.325
15.6 A.P. Roy: Can. J. Phys. *49*, 277 (1971)
15.7 D.K. Ross: J. Phys. C *6*, 3525 (1973)
15.8 K.K. Mani, R. Ramani: Phys. Status Solidi (b) *61*, 659 (1974)
15.9 A.P.P. Nicholson, D.J. Bacon: J. Phys. C *10*, 2295 (1977)
15.10 B.N. Brockhouse, G. Shirane in: *Lattice Dynamics*, ed. by M. Balkanski (Flammarion, Paris 1978) p.123
15.11 N. Wakabayashi, H.G. Smith, R. Shanks: Phys. Lett. A *50*, 367 (1974)
15.12 N. Wakabayashi: Nuovo Cimento B *38*, 256 (1977)
15.13 N. Wakabayashi, H.G. Smith, R.M. Niclow: Phys. Rev. B *12*, 659 (1975)
15.14 J.L. Brebner, S. Jandl, B.M. Powell: Nuovo Cimento B *38*, 263 (1977)
15.15 B. Dorner, R.E. Gosh, G. Harbecke: Phys. Status Solidi (b) *73*, 655 (1976)
15.16 W.G. Stirling, B. Dorner, J.D.N. Cheeke, J. Revelli: Solid State Commun. *18*, 931 (1976)
15.17 D.E. Moncton, F.J. DiSalvo, J.D. Axe in: *Lattice Dynamics*, ed. by M. Balkanski (Flammarion, Paris 1978) p.561
15.18 A. Pasternak: J. Phys. C *9*, 2987 (1976)
15.19 W.B. Yelon, S. Scherm, C. Vettier: Solid State Commun. *15*, 391 (1974)
15.20 A. Anderson, J.P. Todoeschuck: Can. J. Spectrosc. (Canada) *22*, 113 (1977)
15.21 P. Carrara, J.P. Redoules, C. Escribe, K.R.A. Ziebeck: Solid State Commun. *21*, 929 (1977)
15.22 K. Wagner, G. Dolling, B.M. Powell, G. Landwehr: Phys. Status Solidi (b) *85*, 211 (1978)
15.23 H. Meyer, A. Weiss, B. Dorner: Solid State Commun. *25*, 1093 (1978)
15.24 B.M. Powell, S. Jandl. J.L. Brebner, F. Lévy: J. Phys. C *10*, 3039 (1977)
15.25 A. Polian, K. Kunc, R. Le Toullec, B. Dorner in: *Physics of Semiconductors, 1978*, ed. by B.L.H. Wilson (The Institute of Physics, London 1979) p.907
15.26 S. Jandl. J.L. Brebner, B.M. Powell: Phys. Rev. B *13*, 686 (1976)
15.27 W. Kress, A. Frey, B. Dorner, W. Kaiser in: *Proc. Conf. Neutron Scattering*, Vol. 1, ed. by R.M. Moon (Gatlinburg, USA 1976) p.216
15.28 H.J.L. van der Valk: Solid State Commun. *20*, 815 (1976)
15.29 A. Frey, R. Zeyher: Solid State Commun. *28*, 435 (1978)
15.30 D.E. Moncton, J.D. Axe, F.J. DiSalvo: Phys. Rev. B *16*, 801 (1977)
15.31 H.G. Smith, N. Wakabayashi in: *Dynamics of Solids and Liquids by Neutron Scattering*, ed. by T. Springer, S. Lovesey (Springer Berlin, Heidelberg, New York 1977)

16.1 W.D. Teuchert, R. Geick, G. Landwehr, H. Wendel, W. Weber: J. Phys. C 8, 3725 (1975)
16.2 J. Etchepare, P. Kaplan, M. Merian in: *Lattice Dynamics*, ed. by M. Balkanski (Flammarion, Paris 1978) p.60
16.3 B.A. Kotov, N.M. Okuneva, A.R. Regel, A.L. Shakh-Budagov: Sov. Phys. Solid State 9, 955 (1967)
16.4 R. Geick, U. Schröder, J. Stuke: Phys. Status Solidi 24, 99 (1967)
16.5 T. Nakayama, A. Odajima: J. Phys. Soc. Jpn. 34, 732 (1973)
16.6 W. Hamilton, B. Lassier, M. Kay: J. Phys. Chem. Solids 35, 1089 (1974)
16.7 H. Wendel, W. Weber, W.D. Teuchert: J. Phys. C 8, 3737 (1975)
16.8 B.M. Powell, P. Martel: J. Phys. Chem. Solids 36, 1287 (1975)
16.9 W. Glisser, A. Axmann, T. Springer in: *Neutron Inelastic Scattering*, Vol. 1 (IAEA, Vienna 1968) p.245
16.10 S.A. Pine, G. Dresselhaus: Phys. Rev. B 4, 356 (1971)
16.11 E.R. Cowley: Can. J. Phys. 51, 843 (1973)
16.12 T.B. Gibbons: Phys. Rev. B 7, 1410 (1973)
16.13 H. Wendel: J. Phys. C 9, 445 (1976)
16.14 L. Pintschovius, R. Currat, H. Wendel in: *Lattice Dynamics*. ed. by M. Balkanski (Flammarion, Paris 1978) p.579
16.15 H. Wendel: J. Phys. C 10, L1 (1977)
16.16 T. Luthy, G.S. Pawley: Phys. Status Solidi (b) 69, 551 (1975)
16.17 W. Reichardt, K.H. Rieder in: *Proc. Conf. Neutron Scattering*, Vol. 1, ed. by R.M. Moon (Gatlinburg, USA 1976) p.181
16.18 G.A. Briggs, C. Duffill, M.T. Hutchings, R.D. Lowde, N.S. Satya-Murthy in: *Neutron Inelastic Scattering* (IAEA Vienna, 1972) p.669
16.19 M.M. Beg, S.M. Shapiro: Phys. Rev. B 13, 1728 (1976)
16.20 C. Carabatos, B. Prevot: Phys. Status Solidi (b) 44, 701 (1971)
16.21 H. Prask, H. Boutin, S. Yip: J. Chem. Phys. 48, 3367 (1968)
16.22 T.H.K. Barron, C.C. Huang, A. Pasternak: J. Phys. C 9, 3925 (1976)
16.23 H. Grimm, H. Rzany, B. Dorner, H. Jagodzinski: Priv. commun. 1978
16.24 M.M. Elcombe: as cited in Ref. 16.22
16.25 M.M. Elcombe: Proc. Phys. Soc. (London) 91, 947 (1967)
16.26 M.E. Striefler, G.R. Barsch: Phys. Rev. B 12, 4553 (1975)
16.27 P. Tompson, N.W. Grimes: Solid State Commun. 25, 609 (1978)
16.28 M. Iszumi, J.D. Axe, G. Shirane, K. Shimaoka: Phys. Rev. B 15, 4392 (1977)
16.29 R.P. Rinaldi, G.S. Pawley: J. Phys. C 8, 599 (1975)
16.30 B.A. Kortov, N.M. Okuneva, A.L. Shakh-Budagov: Sov. Phys. Solid State 9, 2011 (1968)
16.31 J. Skalyo, B.C. Frazer, G. Shirane: Phys. Rev. B 1, 278 (1970)
16.32 P. Bosi, R. Tubino, G. Zerbi: J. Chem. Phys. 59, 4578 (1973)
16.33 B. Renker: Phys. Lett. A 30, 493 (1969)
17.1 J.K. Kjems, G. Dolling: Phys. Rev. B 11, 1639 (1975)
17.2 A. Pasternak, A. Anderson, J.W. Leech: J. Phys. C 10, 3261 (1977)
17.3 H.G. Smith, M. Nielsen, C.B. Clark: Chem. Phys. Lett. (Netherlands) 33, 75 (1975)
17.4 B.M. Powell, G. Dolling, L. Piseri, P. Martel in: *Neutron Inelastic Scattering* (IAEA, Vienna 1972) p.207
17.5 P.S. Goyal, B.A. Dasannacharya, C.L. Thaper, P.K. Iyengar: Phys. Status Solidi (b) 50, 701 (1972)
17.6 S.F. Trevino, M.K. Farr, P.A. Giguera, J.L. Arman in: *Proc. Conf. Neutron Scattering*, ed. by R.M. Moon (Gatlinburg, USA 1976) p.152
17.7 P. Bruesch, M. Bösch, W. Känzig, M. Ziegler, W. Bührer: Phys. Status Solidi (b) 77, 153 (1976)

17.8 K.R. Rao, S.F. Trevino, H. Prask: Phys. Rev. B *4*, 4551 (1971)
17.9 I.U. Heilmann, N.B. Olsen, D.J. Lockwood, G.A. Mackenzie, G.S. Pawley in: *Lattice Dynamics*, ed. by M. Balkanski (Flammarion, Paris 1978) p.475
17.10 J. Sakurai, R.A. Cowley, G. Dolling: J. Phys. Soc. Jpn. *28*, 1426 (1970)
18.1 L.K. Vodopyanov, I.V. Kutcherenko, A.P. Shotov, R. Scherm in: *Lattice Dynamics*, ed. by M. Balkanski (Flammarion, Paris 1978) p.673
18.2 W.J. Daughten, E. Gürmen, C.W. Tompson in: *Proc. Conf. Neutron Scattering*, Vol. 1, ed. by R.M. Moon, (Gatlinburg, USA 1976) p.145
18.3 C.H. Perry, I.R. Jahn, V. Wagner, W. Bauhofer, L. Genzel, J.B. Sokoloff in: *Lattice Dynamics*, ed. by M. Balkanski (Flammarion, Paris 1978) p.419
18.4 Y. Yamada, Y. Noda, J.D. Axe, G. Shirane: Phys. Rev. B *9*, 4429 (1974)
18.5 L. Genzel, T.P. Martin, C.H. Perry: Phys. Status Solidi (b) *62*, 83 (1974)
18.6 H.G. Smith, J.G. Traylor, W. Reichardt: Phys. Rev. *181*, 1218 (1969)
18.7 W.J.L. Buyers, R.A. Cowley in: *Inelastic Neutron Scattering*, Vol. 1 (IAEA, Vienna 1968) p.43
18.8 J. Asam, S. Rolandson, M.M. Beg. N.M. Butt, Q.H. Khan: Phys. Status Solidi (b) *77*, 693 (1977)
18.9 N.E. Massa, J.F. Vetelino, S.S. Mitra in: *Lattice Dynamics*, ed. by M. Balkanski (Flammarion, Paris 1978) p.425
19.1 W.G. Stirling, W. Press, H.H. Stiller: J. Phys. C *10*, 3959 (1977)
19.2 B. Dorner, H. Stiller: Phys. Status Solidi *18*, 795 (1966)
19.3 H. Stiller, S. Hautecler in: *Inelastic Scattering of Neutrons in Solids and Liquids*, Vol. 2, (IAEA, Vienna 1963) p.281
19.4 B. Dorner, H. Stiller: Phys. Status Solidi *5*, 511 (1964)
19.5 W. Press, B. Dorner, H. Stiller: Solid State Commun. *9*, 1113 (1971)
19.6 P.A. Reynolds: J. Chem. Phys. *59*, 2777 (1973)
19.7 P.A. Reynolds, J.K. Kjems, J.W. White: J. Chem. Phys. *60*, 824 (1974)
19.8 G. Dolling, B.M. Powell: Proc. Roy. Soc. A *319*, 209 (1970)
19.9 L.N. Becka: J. Chem. Phys. *37*, 431 (1962)
19.10 W. Bührer, W. Hälg, T. Schneider in: Rep. Inst. Reaktorforsch., AF-SSP-9, Würenlingen 1967
19.11 G. Dolling, B.M. Powell, G.S. Pawley: Proc. Roy. Soc. A *333*, 368 (1973)
19.12 T. Wasiutynski: Phys. Status Solidi (b) *76*, 175 (1976)
19.13 P. Martel, B.M. Powell, L.A. Vinhas in: *Lattice Dynamics*, ed. by M. Balkanski, (Flammarion, Paris 1978) p.494
19.14 P. Martel, B.M. Powell: Chem. Phys. Lett. *39*, 339 (1976)
19.15 E.L. Bokhenkov, E.F. Sheka, B. Dorner, I. Nathaniec: Solid State Commun. *23*, 89 (1977)
19.16 E.L. Bokhenkov, E.F. Sheka, I. Nathaniec, B. Dorner, W. Drexel in: *Lattice Dynamics*, ed. by M. Balkanski, (Flammarion, Paris 1978) p.471
19.17 A.I. Kitaigorodski: J. Chem. Phys. *63*, 9 (1966)
19.18 D.E. William: J. Chem. Phys. *45*, 3370 (1966); *47*, 4680 (1967)
19.19 G.S. Pawley, E.A. Yeats: Solid State Commun. *7*, 385 (1969)
19.20 E.L. Bokhenkov, I. Nathaniec, E.F. Sheka: Sov. Phys. JETP *43*, 536 (1976)
19.21 G.A. Mackenzie, G.S. Pawley, O.W. Dietrich: J. Phys. C *10*, 3723 (1977)
19.22 U.A. Lutz, W. Halg: Solid State Commun. *8*, 165 (1970)
19.23 J. Lefebvre, M. More, R. Fouret, B. Hennion, R. Currat: J. Phys. C *8*, 2011 (1975)

Subject Index

Abbreviations used in figure captions 15
ABO_3 crystals 141-159
ABX_3 crystals 141-159
Adiabatic approximation 6
Adiabatic condition 9
AgBr 12,80
AgCl 79
AgI (β-) 121
Alkali halides 8,12,27-47
Aluminates:
    lathanum aluminate 153
    magnesium aluminate 189
Aluminum antimonide 103
AlSb 103
Ammonia 197
Ammonium bromide 91
Ammonium bromide / Ammonium chloride 206
Ammonium chloride 89,90
Ammonium iodide 83
Anharmonicity 6
Anomaly, phonon 11
Anthracene 217
Antimonides:
    aluminum antimonide 103
    gallium antimonide 106
    indium antimonide 109
    neodymium antimonide 78
Approximation, adiabatic 6,9
Approximation, harmonic 6

Ar 24
Argon 24
As (α-) 184
Arsenic 184
Arsenides:
    gallium arsenide 105
    indium arsenide 108
ASM 15
Axially symmetric model 15
Azide, potassium 200

BaF_2 127
BaO 53
Barium fluoride 127
Basu 11
$BaLiH_3$ 152
Barium lithium hydride 152
Barium oxide 53
$BaTiO_3$ (cubic) 149
$BaTiO_3$ (tetragonal) 150,151
Barium titanate 149-151
BCM 12
Benzene, dichloro- 212
BeO 118
Beryllium oxide 118
Bismuth telluride 171
Bi_2Te_3 171
Bilz 14
Bond angle 13
Bond charge model 12,15
Bond distance 13

Born-Mayer potential 7
Born model 7
Boundary conditions 6
Breathing deformability 11,13
Breathing shell model 11,13,15
Bromides:
    ammonium bromide 91,206
    cesium bromide 87
    copper bromide 116
    lithium bromide 32
    potassium bromide 41,42,207
    rubidium bromide 46
    silver bromide 80
    sodium bromide 37
    strontium bromide 126
    thallium bromide 93
BSM 11,13

C (diamond) 96
C (graphite) 162,163
$CaCO_3$ 157
Cadmium sulfide 120
Cadmium telluride 113
CaF_2 124
Calcium carbonate 157
Calcium fluoride 124
Calcium oxide 51
CaO 51
Carbides:
    hafnium carbide 65
    niobium carbide 67,68
    silicon carbide 102
    tantalum carbide 70
    titanium carbide 61
    uranium carbide 71
    vanadium carbide 66
    zirconium carbide 63
Cabon dioxide 196
Cauchy relation 13

CD_4 210
$C_{10}D_8$ 216
$C_{14}D_{10}$ 217
$C_6D_4Cl_2$ 212
$C_6D_{12}N_4$ 213
$C_4D_5ON_3$ 215
$C_5D_6O_2N_2$ 214
$CO(ND_2)_2$ 218
CdS 120
CdTe 113
Cesium bromide 87
Cesium chloride 86
Cesium chloride structure crystals 85-93
Cesium fluoride 84
Cesium iodide 88
Cesium nickel fluoride 155
Charge 7-14
Charge density 14
Charge density, electronic 13
Charge distortion 8,11
Charge, electronic 9,12
Charge, ionic 7,9
Charge, overlap 10
Charge, Szigeti 9
Charge, transfer 11,12
Chlorides:
    ammonium chloride 89,90,206
    cesium chloride 86
    cobaltous chloride 170
    copper chloride 114
    ferrous chloride 169
    lithium chloride 31
    potassium chloride 40
    rubidium chloride 45
    silver chloride 79
    sodium chloride 36
    strontium chloride 126
    thallium chloride 92

$C_4N_2D_4$ 211
$CoCl_2$ 170
CO_2 196
Cobaltous chloride   170
Cobaltous fluoride   139
Cobaltous oxide   56
Cobaltous potassium fluoride   156
Cochran   8,10
CoF_2 139
$Co(ND_2)_2$ 218
Condition, adiabatic   9
Conservation laws   6
Constants, dielectric   3
Constants, elastic   3
Constants, force   3,6,7
Conversion factors for phonon energies   16
CoO (antiferromagnetic)   57
CoO (paramagnetic)   56
Copper bromide   115
Copper chloride   114
Copper halides   12,114-116
Copper iodide   116
Copper oxide (red)   186
Core electrons   14
Correlation   14
Coulomb forces   8,14
Coulomb interaction   7
Coulomb matrix   8
Coupling, electron-electron   9
Coupling, electron-ion   9
Coupling, core-shell   9
Covalent crystals   12,13
Cuprite   186
CsBr   87
CsCl   86
CsF   84
CsI   88
$CsNiF_3$ 115

CuBr   115
CuCl   114
CuI   116
Cu_2O 186
Cyanides:
　sodium cyanide   81
　potassium cyanide   82
Cytosine   215

DBM   15
DDM   10
Deformable bond model   15
Deformable shell model   11,13,15
Deformation, breathing   11,13
Deformation, dipolar   11
Deformation dipole model   15
Deformation, monopolar   11
Deformation, quadrupolar   11-13
DEFSM   11
Density of states   7
Deuteride, lithium   28,29
Diamond   12,13,96
Diamond structure crystals   95-99
Dichlorobenzene   212
Dick   8
Dielectric constants   3
Dielectric function   14
Dioxides:
　carbon dioxide   196
　silicon dioxide   188
　sodium dioxide   199
　titanium dioxide   134
　uranium dioxide   131
Dipolar deformation   11
Dipolar forces   9
Dipole model   8,14
Dispersion relation   3,6
Displaced ion   8
Displacement vector   6

Disulfides:
    molybdenum disulfide   165
Diselenides:
    niobium diselenide   164
    tantalum diselenide   177
    titanium diselenide   168
DM   8
D_2O (ice, hexagonal)   187
D_2O_2 198
Double shell model   11,15
D-p-hybridization   12
DSM   11
Dynamical matrix   6,8

Eigenvalue   6
Eigenvector   6,7
Elastic constants   3
Electron, core   14
Electron, valence   14
Electronic charge   9,12
Electronic charge density   13
Electronic polarization   8
Electronic screening   8
Electron-electron interaction   14
Electron-electron coupling   9
Electron-ion coupling   9
Electron-ion interaction   14
Energy, free   6
Energy, lattice   7
Equation of motion   6,8
Equilibrium condition   8
Equilibrium position   6
ESM   10
Extended shell model   10
Exchange   14

FCM   6
$FeCl_2$ 169
FeF_2 138

FeO   55
Ferrous chloride   169
Ferrous fluoride   138
Ferrous oxide   55
Figure captions, abbreviations used in   15
Fischer   12
Fluorides:
    barium fluoride   127
    calcium fluoride   124
    cesium fluoride   84
    cesium nickel fluoride   155
    cobaltous fluoride   139
    ferrous fluoride   138
    lead fluoride   128
    lithium fluoride   30
    magnesium fluoride   136
    manganous fluoride   137
    manganous potassium fluoride   146
    potassium cobaltous fluoride   156
    potassium fluoride   39
    potassium manganous fluoride   146
    rubidium fluoride   44
    sodium fluoride   35
    strontium fluoride   125
Fluorite   124
Fluorite structure crystals   123-131
Force, Coulomb   8,14
Force constant model   15
Force constants   3,6,7
Forces, dipolar   9
Forces, overlap   7
Forces, valence   12,13
Free energy   6

GaAs   105
Gallium antimonide   106
Gallium arsenide   105
Gallium phosphide   104

Gallium selenide   175
Gallium sulfide   173,174
GaP   104
GaS   173,174
GaSb   106
GaSe   175
Ge   98
Germanium   12,13,98
Germanium sulfide   176
GeS   176
Graphite   162,163

Hafnium carbide   65
HfC   65
Halides, alkali   8,12,27-47
Halides, copper   12,114-116
Halides, silver   12,79,80,121
Hanke   14
Harmonic approximation   6
He (bcc)   22
He (fcc)   20
He (hcp)   21
Helium (bcc)   22
Helium (fcc)   20
Helium (hcp)   21
Hexa-methylen-tetramine   213
Hybridization   13
Hydrides:
    barium lithium hydride   152
    lithium hydride   28,29
    sodium hydride   34
    strontium lithium hydride   148
Hydrogen peroxide   198

I_2 195
Ice   187
InAs   108
Indium antimonide   109
Indium arsenide   108

Indium phosphide   107
Infrared spectra   3
InP   107
InSb   109
Interaction, Coulomb   7
Interaction, electron-electron   14
Interaction, electron-ion   14
Internal displacement coordinates   12
Iodides:
    ammonium iodide   83
    cesium iodide   88
    copper iodide   116
    lead iodide   167
    lithium iodide   33
    potassium iodide   43
    potassium rubidium iodide   208
    rubidium iodide   47
    silver iodide   121
    stannic iodide   201
Iodine   195
Ionic charge   7,9
Ionic crystals   7

KBr   41,42
$(KBr)_{1-x}(RbBr)_x$ 207
KCl   40
KCN   82
$KCoF_3$ 156
KD_2PO_4 191
Keating   12
Kellermann model   8
KF   39
KI   43
$KMnF_3$ 146
KN_3 159
Kr   25
$K_{0.5}Rb_{0.5}I$ 208
Krypton   25

K_2SeO_4 190
$KTaO_3$ 144,145
Kunc   13

$LaAlO_3$ 153
Lanthanum aluminate   153
Lattice energy   7
Layered structure crystals   161-177
Lead fluoride   128
Lead iodide   167
Lead magnesium   130
Lead selenide   76
Lead sulfide   75
Lead telluride   77
Lead titanate   154
Lead tin selenide   204
Lead tin telluride   205
Lennard-Jones potential   19
LiD   14,28,29
LiBr   32
LiCl   31
LiF   30
LiH   28,29
LiI   33
$LiNbO_3$ 142
Lithium barium hydride   152
Lithium bromide   32
Lithium chloride   31
Lithium deuteride   28,29
Lithium fluoride   30
Lithium hydride   28,29
Lithium iodide   33
Lithium niobate   142
Lithium strontium hydride   148
Localized wave functions   14
Lyddane-Sachs-Teller relation   8

Madelung constant   7
Magnesium aluminate   189
Magnesium fluoride   136
Magnesium lead   130
Magnesium oxide   50
Magnesium stanide   129
Manganous fluoride   137
Manganous oxide   54
Manganous potassium fluoride   146
MASM   15
Mass, reduced   9
Matrix, Coulomb   8
Matrix, dynamical   6,8
Metal oxides   49-57
Methane   210
Methylen-tetramin, hexa-   213
Methyl-thymine   214
$MgAl_2O_4$ 189
MgF_2 137
MgO   50
Mg_2Pb 130
Mg_2Sn 129
Microscopic theory   14
Mixed crystals   203-208
MnF_2 137
Models:
   Born model   7
   bond charge model   12,15
   breathing shell model   11,13,15
   deformable bond model   15
   deformable shell model   11,13,15
   dipole model   8,14
   double shell model   11,15
   extended shell model   10
   force constant model   15
   Kellermann model   8
   overlap shell model   10,15
   quadrupole shell model   11
   rigid ion model   7,15
   rigid layer model   15
   shell model   8-10,15

three body force shell model
  13,15
valence field model 12,13
valence force deformation dipole
  model 15
valence force shell model 15
valence force overlap shell model
  13
Modified axially symmetric model 15
Modified shell model 15
Molecular crystals 193-202
Molecular methode 13
Molybdenite 165
Molybdenum disulfide 165
Monopolar deformation 11
MoS_2 165
MnO 54
MSM 15

N_2 (α-) 194
NaBr 37
NbC 67,68
NaCl 36
NaCN 81
NaF 35
NaH 34
NaI 38
$NaNO_2$ 202
$NaNO_3$ 158
NaO_2 199
Na_2S 123
NbN 69
$NbSe_2$ 164
ND_4Cl 90
ND_4Br 91
ND_4I 83
NdSb 78
NH_3 197

NH_4Cl 89
$NH_4Cl_{1-x}Br_x$ 206
Naphtalene 216
Ne 23
Neon 23
Neutron scattering 3
Neodymium antimonide 78
Nickel cesium fluoride 155
Nickel oxide 57
Nickel sulfide 185
NiO 58
Niobates:
  lithium niobate 142
  potassium niobate 143
Niobium carbide 67,68
Niobium diselenide 164
Niobium nitride 69
NiS 185
Nitrates:
  potassium nitrate 159
  sodium nitrate 158,202
Nitrides:
  niobium nitride 69
  poly-sulfur nitride 182
  titanium nitride 62
  zirconium nitride 64
  uranium nitride 72
Nitrogen (α-) 194

Optical constants 9
Optic modes 8
Organic crystals 209-218
OSM 13
Overhauser 8
Overlap 14
Overlap charge 10
Overlap force 7
Overlap shell model 10,15

Oxides:
    barium oxide   53
    beryllium oxide   118
    calcium oxide   51
    cobaltous oxide   56
    copper oxide   186
    ferrous oxide   55
    manganous oxide   54
    magnesium oxide   50
    nickel oxide   58
    stannic oxide   140
    strontium oxide   52
    zinc oxide   119

Paraffin   13
PbF_2 128
PbI_2 167
PbS   75
PbSe   76
$Pb_{0.8}Sn_{0.2}Se$ 204
$Pb_{0.87}Sn_{0.13}Te$ 205
$PbTiO_3$ 154
PbTe   77
Perovskite-like compounds:
    $BaLiH_3$ 152
    $BaTiO_3$ 149-151
    $CaCO_3$ 157
    $KCoF_3$ 156
    $KMnF_3$ 146
    $KNbO_3$ 143
    $KTaO_3$ 144,145
    $LaAlO_3$ 153
    PbTiO   154
    $SrLiH_3$ 148
    $SrTiO_3$ 147
Phonon anomaly   11
Phosphates
    potassium phosphate   191

Phosphides:
    gallium phosphide   104
    indium phosphide   107
Polarizability   9
Polarizability electronic   8
Polarization   10
Polarization, electronic   8
Poly-sulfur nitride   182
Potassium azide   200
Potassium bromide   41,42
Potassium chloride   40
Potassium cobaltous fluoride   156
Potassium cyanide   82
Potassium fluoride   39
Potassium iodide   43
Potassium manganous fluoride   146
Potassium niobate   143
Potassium nitrate   159
Potassium phosphate   191
Potassium rubidium bromide   207
Potassium rubidium iodide   208
Potassium selenate   190
Potassium tantalate   144,145
Potential   6
Potential, Born - Mayer   7
Potential, Lennard - Jones   19
Pseudopotential   14
Pyrazine   211

QSM   11
Quadrupolar deformation   11-13
Quadrupolar force   12
Quadrupole shell model   11
Quantum crystals   19
Quartz (α-)   188

Raman spectra   3
Rare gas crystals   19-26
RbBr   46

RbCl 45
RbF 44
RbI 47
Reduced mass 9
Representation, symmetry 11
Rigid ion model 7,15
Rigid layer model 15
RIM 7
RLM 15
Rock salt 36
Rock salt structure crystals 27-84
Rubidium bromide 46
Rubidium chloride 45
Rubidium fluoride 44
Rubidium iodide 47
Rubidium potassium bromide 207
Rubidium potassium iodide 208
Rutile 134,135
Rutile structure crystals 133-140

S_8 183
Samarium sulfide 73
Samarium yttrium sulfide 73
Scattering, neutron 3
Schachtschneider 13
Schröder 11
Screening, electronic 8
Se 180
Selenates:
    potassium selenate 190
Selenides:
    gallium selenide 175
    lead selenide 76
    lead tin selenide 204
    zinc selenide 76
Selenium 180
Sham 14
Shell 9
Shell-core coupling 9

Shell model 8-10,15
Si 97
SiC 102
Silica 188
Silicon 12,13,97
Silicon carbide 102
Silicon dioxide 188
Singh 13
Sinha 14
SiO_2 188
Silver bromide 80
Silver chloride 79
Silver halides 12,79,80,121
Silver iodide 121
SmS 73
SM 8
$Sm_{0.75}Y_{0.25}S$ 73
Sn (α-) 99
$(SN)_x$ 182
SnI_4 201
SnO_2 140
$SnSe_2$ 166
SnTe 74
Snyder 13
Sodium bromide 37
Sodium chloride 36
Sodium cyanide 81
Sodium dioxide 199
Sodium fluoride 35
Sodium hydride 34
Sodium iodide 38
Sodium nitrate 158
Sodium nitrite 202
Sodium sulfide 123
Space group 6
Spectra, infrared 3
Spectra, Raman 3
Stability 6
Stanide, magnesium 129

Stannic iodide 201
Stannic oxide 140
Stannic selenide 166
Stannous telluride 74
Strontium chloride 126
Strontium fluoride 125
Strontium gallium 172
Strontium lithium hydride 148
Strontium oxide 52
Strontium titanate 147
$SrCl_2$ 126
SrF_2 125
SrGa 172
$SrLiH_3$ 148
SrO 52
$SrTiO_3$ 147
Sulfides:
    cadmium sulfide 120
    gallium sulfide 173,174
    germanium sulfide 176
    lead sulfide 75
    nickel sulfide 185
    yttrium sulfide 60
    zinc sulfide 110
Sulfur 183
Sulfur nitride, poly- 182
Superconductors 11
Symmetry coordinates 13
Symmetry representations 11
Symmetry restrictions 6
Szigeti charge 9

TaC 70
Tantalates:
    potassium tantalate 144,145
Tantalum carbide 70
Tantalum diselenide 177
TaS_2 177
Te 181

Tellurium 181
Tellurides:
    bismuth telluride 171
    cadmium telluride 113
    lead telluride 77
    lead tin telluride 205
    stannous telluride 74
    zinc telluride 112
Tetramin, hexa-methylen- 213
Thallium bromide 93
Thallium chloride 92
Thermal expansion 6
Three-body force shell model 13,15
Thymine, methyl- 214
TiC 61
TiN 62
Tin (α-) 99
Tin (gray) 13,99
Tin lead selenide 204
Tin lead telluride 205
TiO_2 134,135
$TiSe_2$ 168
Titanates:
    barium titanate 149-151
    lead titanate 154
    strontium titanate 146
Titanium carbide 61
Titanium dioxide 134,135
Titanium diselenide 168
Titanium nitride 62
TlBr 93
TlCl 92
Transition metal compounds 59-72
TSM 13

UC 71
UN 72
UO_2 131
Uranium carbide 71

Uranium dioxide   131
Uranium nitride   72
Urea   218

Valence electrons   14
Valence forces   12,13
Valence force deformation dipole model   15
Valence force field   10
Valence force field model   12,13
Valence force model   15
Valence force shell model   15
Valence overlap shell model   13,15
Vanadium carbide   66
VC   66
VDDM   15
Verma   13
VFFM   12,13
VOSM   13
VSM   15

Wave functions, localized   14
Weber   11,12
Wurtzite structure crystals   117-121

Xe   26
Xenon   26

YS   60
Yttrium sulfide   60

Zeyher   10,13
Zinc blende   119
Zinc blende structure   13
Zinc blende structure crystals   101-121
Zinc oxide   119
Zinc selenide   13,111
Zinc sulfide   110
Zinc telluride   112
Zirconium carbide   63
Zirconium nitride   64
ZnO   119
ZnS   110
ZnSe   13,111
ZnTe   112
ZrC   63
ZrN   64

J. L. Birman
Licht und Materie I b
Light and Matter I b
Editor: L. Genzel
1974. 34 figures. XVI, 538 pages
(Handbuch der Physik, Gruppe 5,
25. Band, 2. Teil b)
ISBN 3-540-06638-1

"... The choice of what is important and the development of the theme from the general to the very particular suited me admirably, and I feel sure that solid-state physicists will benefit greatly from this part."
Contemporary Physics

"...an excellent article of encyclopedia..."
Physics Bulletin

Light Scattering in Solids
Editor: M. Cardona
1975. 111 figures, 3 tables. XIII, 339 pages
(Topics in Applied Physics, Volume 8)
ISBN 3-540-07354-X

Contents:
M. Cardona: Introduction. – *A. Pinczuk: E. Burstein:* Fundamentals of Inelastic Light Scattering in Semiconductors and Insulators. – *R. M. Martin, L. M. Falicov:* Resonant Raman Scattering. – *M. V. Klein:* Electronic Raman Scattering. – *M. H. Brodsky:* Raman Scattering in Amorphous Semiconductors. – *A. S. Pine:* Brillouin Scattering in Semiconductors. – *Y.-R. Shen:* Stimultated Raman Scattering

Photoemission in Solids I
General Principles
Editors: M. Cardona, L. Ley
1978. 90 figures, 17 tables. XI, 290 pages
(Topics in Applied Physics, Volume 26)
ISBN 3-540-08685-4

Contents:
M. Cardona, L. Ley: Introduction. – *W. L. Schaich:* Theory of Photoemission: Independent Particle Model. – *S. T. Manson:* The Calculation of Photoionization Cross Sections: An Atomic View. – *D. A. Shirley:* Many-Electron and Final State Effects: Beyond the One-Electron Picture. – *G. K. Wertheim, P. H. Citrin:* Fermi Surface Excitations in X-Ray Photoemission Line Shapes From Metals. – *N. V. Smith:* Angular Dependent Photoemission.

Photoemission in Solids II
Case Studies
Editors: L. Ley, M. Cardona
1979. 214 figures, 26 tables.
XVIII, 401 pages
(Topics in Applied Physics, Volume 27)
ISBN 3-540-09202-1

Contents:
L. Ley, M. Cardona: Introduction. – *L. Ley, M. Cardona, R. A. Pollak:* Photoemission in Semiconductors. – *S. Hüfner:* Unfilled Inner Shells: Transition Metals and Compounds. – *M. Campagna, G. K. Wertheim, Y. Baer:* Unfilled Inner Shells: Rare Earths and Their Compounds. – *W. D. Grobman, E. E. Koch:* Photoemission from Organic Molecular Crystals. – *C. Kunz:* Synchrotron Radiation: Overview. – *P. Steiner, H. Höchst, S. Hüfner:* Simple Metals.

Springer-Verlag
Berlin
Heidelberg
New York

W. Ludwig
Recent Developments in Lattice Theory

1967. 87 figures. VI, 301 pages
(Springer Tracts in Modern Physics, Volume 43)
ISBN 3-540-03982-1

Contents:
Introduction. – Symmetry and Invariance Properties of the Coupling Parameters. – Different Models for the Potential Energy in Crystals. – Dynamics of Molecular Crystals. – Anharmonic Effects in Thermodynamical Properties. – Point Defects in Crystal Lattices. – Interaction of Phonons with Particles and Radiation. – Appendix. – References.

Excitions

Editor: K. Cho
1979. 118 figures, 4 tables. Approx. 290 pages
(Topics in Current Physics, Volume 14)
ISBN 3-540-09567-5

Contents:
K. Cho: Introduction. – *K. Cho:* Internal Structure of Excitions. – *P. J. Dean, D. C. Herbert:* Bound Excitions in Semiconductors. – *B. Fischer, J. Lagois:* Surface Excition Polaritons. – *P. Y. Yu:* Study of Excitions and Excition-Phonon Interactions by Resonant Raman and Brillouin Spectroscopies.

G. Leibfried, N. Breuer
Point Defects in Metals I

Introduction to the Theory

1977. 138 figures, 22 tables. XIV, 342 pages
(Springer Tracts in Modern Physics, Volume 81)
ISBN 3-540-08375-8

Contents:
Introduction and Survey. – Harmonic Approximation and Linear Response (Green's Function) of an Arbitrary System. – Lattice Theory. – Continuum Theory. – Transition From Lattice to Continuum Theory. – Statistics and Dynamics of Simple Single Point Defects. – Scattering of Neutrons and X-Rays by Crystals. – Probability, Distributions and Statistics. – Properties of Crystals With Defects in Small Concentration. – Appendix.

Springer-Verlag
Berlin
Heidelberg
New York

QC
176.8
R3
B54

JAN 15 1980